Tutorial Distance Learning
Rebuilding Our Educational System

INNOVATIONS IN SCIENCE EDUCATION AND TECHNOLOGY

Series Editor:

Karen C. Cohen, Harvard University, Cambridge, Massachusetts

A Continuation Order Plan is available for this series. A continuation order will bring delivery of each new volume immediately upon publication. Volumes are billed only upon actual shipment. For further information please contact the publisher.

Tutorial Distance Learning
Rebuilding Our Educational System

Alfred Bork

University of California, Irvine
Irvine, California

Sigrun Gunnarsdottir

Iceland Telecom
Reykjavik, Iceland

Kluwer Academic/Plenum Publishers
New York, Boston, Dordrecht, London, Moscow

Library of Congress Cataloging-in-Publication Data

Bork, Alfred and Sigrun Gunnarsdottir
Tutorial distance learning: rebuilding our educational system/Alfred Bork and Sigrun
Gunnarsdottir.
    p.   cm. — (Innovations in science education and technology)
Includes bibliographical references and index.
ISBN 0-306-46644-9
   1.   2.

LC
5803
.L43
B67
2001

ISBN: 0-306-46644-9

©2001 Kluwer Academic/Plenum Publishers, New York
233 Spring Street, New York, N.Y. 10013

http://www.wkap.nl/

10 9 8 7 6 5 4 3 2 1

A C.I.P. record for this book is available from the Library of Congress

Printed in the United States of America

Contents

Chapter 1

INTRODUCTION

> If we do not invest all our resources of energy and
> will in education, the race with catastrophe will be lost
> and the balance between man and nature will be re-
> established by disasters that are not only unthinkable,
> but also avoidable. The choice is ours, and the time for
> action is now.

Frederico Mayor, Director General of UNESCO
Education for All - Summit of Nine High-Population Countries
New Delhi, 12-16 December 1993
UNESCO Paris France 1994

We are at a critical moment in the history of human learning, and
in the history of humans. Learning is vitally important in our future.
With six billion people on earth, our current educational systems
everywhere at all levels have major problems, probably not curable
with present approaches.

In many areas very little education is available, and where it is
plentiful it is all too often inadequate. Our current learning systems
are weak. We need new learning modes and structures, and we need
them quickly and globally.

This book presents a new system for learning, a system not
reflecting current practice. We believe that it has great potential for
the future of education at all levels worldwide. The learning units in
this system will be available everywhere worldwide, at any time, at a
cost-effective rate for each student.

This approach is based on a new paradigm for learning, computer-
based tutorial learning. The most common delivery system will be

through a new form of distance learning, giving a new meaning to this important concept. Current technology is adequate for this system.

Interaction between the students, singly or in small groups, and the computer, will be very frequent and through the student's native language, such as English or Chinese, similar to that between a student and a human tutor. All these details and others are explained further in the chapters that follow.

We emphasize the word 'potential' in describing the new system. Additional experimental work should explore these ideas, to demonstrate the full effectiveness of the approach suggested. The current problems of learning are great, and, we believe as indicated, not solvable by the approaches currently being pursued. So we strongly urge that this empirical data be carefully gathered and evaluated with this new approach.

Creating new learning materials and systems will require great imagination and determination on our part. We need new ideas that go beyond the learning situations of today, and the ability to develop the necessary materials and institutes. Learning should be greatly improved, less expensive, and available to all in all subjects from birth to death.

Given these serious problems, not enough discussion occurs on desirable futures for learning. The following factors for learning might be considered.

1. Provide a more personalized individualized learning environment for all students. Many lectures offer little individualized help for students, so need to be replaced by individualized approaches. These units must all be developed, as very few exist today.

Highly interactive material, almost conversational in nature, will respond to individual student problems. Learning will be an active process for all students, so that knowledge can be constructed directly by the students.

2. Verify that more students learn and learn better than today, in a shorter time and with better retention. This requires careful experiments with large numbers of students. A major increase in student learning will be attained and verified by large-scale summative and formative evaluations, so that the learning gains are empirically documented.

3. Significantly reduce the costs of education, now too expensive for many students. Interactive technology allows us to reduce costs.

The traditional model of one professor teaching 25 students per class, three times a week . . . will not scale to meet this increased demand because it is too expensive

Carol Twigg, 'The Value of Independent Study'
EDUCOM Review, July-August (1995).

4. Promote research in learning. We must improve our major product, student learning. This should become a primary aim of research.

5. Develop full new modules for many areas. This new learning material will use computers and media. The content will be fully revised to meet modern standards. New modes of student assessment will be part of this development.

6. Market these materials. The units will be marketed widely to institutions and to the general public. Eventually these units should not only recoup the financial costs that have gone into them, but could become major profit sources. We will develop in languages other than English to allow worldwide use.

7. Use these segments as the basis for new distance learning institutions. We would thus greatly increase worldwide student access to learning. Eventually through these efforts we will extend lifelong learning components and work towards an educational system covering everyone from birth to death.

We need a society that focuses on learning, where everyone loves to learn. Solving many of the world's major problems, including population, water, violence, health, and environmental destruction will depend to a major extent on improving learning.

A school or university may undertake this development perhaps in partnership with other schools and universities or commercial partners or these efforts might be entirely commercial.

1.1 Summary of each chapter

We realize that not all readers will be interested in all the details in this book. The following brief summaries of each Chapter may help you to determine what you want to read carefully, at least initially, and what you want to scan. We use some repetition, so those chapters are partially independence.

1.1.1 Problems and vision of learning

We start with the problems of education today, Chapter 2. This will suggest a collection of visions for the future of learning, the basis for the remainder of the book.

1.1.2 Factors of distance learning

The approach proposed in this book assumes distance learning or e-learning which is the fashonable term these days. Many forms of distance learning are possible. Chapter 3 reviews the variables in distance learning, and shows many of the types of distance learning that are possible.

1.1.3 Distance learning – present and future

In Chapter 4 we consider examples of distance learning, using the factors discussed in Chapter 3. Some of these are important historical examples, such as the United Kingdom Open University. Others are imaginative possibilities found in fiction, making interesting and informative suggestions for the future of distance learning.

1.1.4 Individualization and Interaction

We consider, in Chapter 5, first the importance of individualization in learning. Then we look at student computer interactions, with individualization in mind. We develop a methodology for highly interactive learning, considering both frequency and quality of interaction.

1.1.5 Multimedia in Learning

Many media are possible in computer-based tutorial learning units. They will help and motivate many students. In chapter 6 we review here the use of pictures, video, sound (in both directions), and computers.

1.1.6 Tutorial learning

We have mentioned several times earlier the suggested new paradigm for learning. In Chapter 7 we explore the concept of computer-based tutorial learning, using the ideas of interaction just developed.

1.1.7 Delivery of learning

Learning units must reach students in many locations in distance learning. In Chapter 8 we look at various delivery methods that are available, some in common use. Satellites seem particularly important for the future.

1.1.8 Learning and assessment

Learning and assessment are often separated, considered different activities. In computer-based learning they can be combined. This implies that assessment is not used for assigning grades, not needed in a mastery environment, but is used to determine what learning material is to be presented next. We discuss the way to include assessment into learning in Chapter 9.

1.1.9 Structures for technology based learning

New technology in learning suggests new organizational structures for learning material. Particularly important are ways to organize learning that stress mastery and experiential learning, we discuss these issues in chapter 10.

1.1.10 Developing tutorial learing units

Chapter 11 presents the system developed at the University of California, Irvine, and the University of Geneva for developing highly interactive tutorial computer-based learning. Management, design, implementation, and evaluation are considered.

1.1.11 Cost of highly interactive tutorial learning

In Chapter 12 we investigate the important considerations concerning the cost of learning, focusing on the approach considered in this book and comparing it to some extent with other educational strategies. We argue that computer-based tutorial learning is a cost-effective method for the future of learning.

1.1.12 Initial tutorial learning units

We review, in Chapter 13, the possibilities for development, in all areas. The emphasis is on what should be developed first, in terms of both market and need.

1.1.13 Starting new distance learning institution

In this final Chapter (14) we consider the problems in starting new learning programs of the type discussed in this book. Both conventional institutions and new ones, possibly for profit, are considered.

1.1.14 Bibliography

We list the references used in preparing this book.

Chapter 2

PROBLEMS AND VISIONS OF LEARNING

In a passage from Lewis Carroll, Alice is seeking directions from a cat.

> 'Would you tell me please, which way I ought to go from here?' 'That depends a great deal on where you want to go to.' said the Cat. 'I don't much care where.' said Alice. 'Then it doesn't matter which way you go.' said the Cat.

Alice in Wonderland
1960, page 62

Alice's situation, not knowing where she wants to go, and so not having any reasonable choice of how to start out, is common in education today. It is a major problem in planning for the future of education. We need to know our direction for learning. This chapter attempts to give the 'educational Alice' some direction to head toward, a vision for the future.

The question we ask is "What do we want learning to accomplish, both for the individual and the world?" Having such a vision for the beginning of a new millennium is critical, since we need a new direction.

We first begin with some problems in learning today. Then we will look at some visions for the future of learning, based on our consideration of these problems.

It should be obvious at this point that we need a completely new plan and vision for worldwide access to education at all levels, if the next century is to see a quantum leap in educational opportunities everywhere. Obviously this is a great challenge and no easy task.

Theodore Hesburgh
Looking Forward, ed. John Templeton
HarperCollins New York 1993

2.1 The problems of education today

Learning has many major problems. In this section we discuss the major factors that we think are important.

2.1.1 Too many people

Perhaps the major problem for humans is that the world population has increased from two billion in 1927 to six billion people in 1999. Population is still increasing. We have too many people on earth. The *World Population Profile: 1999* (United States Census Bureau) estimates that we will have over nine billion people on earth in 2050.

This large population brings us many problems. Our concern in this book is with learning problems. We have far more people to educate today than we have ever had.

Many of the other problems mentioned are affected by the large number of people who need to learn. Our existing systems will not handle such numbers. They were developed for far fewer people than we have now. Life long learning has also grown dramatically. Although the popularion would not grow at the samt rate it has, the number of learners will.

2.1.2 Many students do not learn

In the developed countries many people do not learn, as seen by our grading systems; grades other than the highest grade indicate incomplete learning. In the poor parts of the world, only limited education is available for most people. All too often no learning is available, except that gained by daily living.

> Schools in developing countries face problems of relatively low school participation in terms of enrollments of eligible age groups, low levels of school completion, even at the primary level, and low levels of achievement . . . their lack of effectiveness is not a mystery, for resources sufficient to provide even the most rudimentary conditions for success often are lacking.

> Marlaine E Lockheed and Henry M Levin
> Effective Schools in Developing Countries
> Falmer Press, London, UK, 1993.
> Copyright by The World Bank

We have a billion illiterate adults in the world, two thirds of them women. This is intolerable; reading is often the key to other learning. Further, this learning shortage is a major handicap in trying to solve the major problems of our earth, such as population, water, violence, and poverty. Although these are often considered to be economic or technological problems, learning is a major ingredient in seeking solutions.

2.1.3 Learning is seldom individualized

A major flaw in lecture-textbook courses, the common strategy found in schools and universities, is that students are mostly treated in the same way, with only minor attempts at individualization. Learning proceeds in lock-step fashion, with all students moving at the same rate and with little attempt to accommodate students who need more time, perhaps to pick up background they are missing or for many other reasons. Most students use the same learning approaches, even if it is ineffective for some students.

Students seldom get much individual attention. This is not surprising, given the numbers of students in the courses and the teaching strategies involved, usually lectures, video, or textbooks.

Often large university introductory course lectures are not taught by the tenure-track faculty, but either partially or totally, by graduate students or by lecturers brought in especially for the courses. As universities continue to lose financial resources, or increase numbers without adequate additional funding, this situation grows worse.

Schools and training have the same problems. The numbers in classes assure that students receive little individual attention to their learning problems. We have come in the United States a long way from the little red schoolhouses, where each student worked at an individualized pace and on material unique to the student. The pressure of large numbers of students is one factor responsible for this change.

2.1.4 Most learning is not active learning.

Another major problem with contemporary educational approaches is that a large amount of learning is passive for most students. Listening to a lecture, or reading from a book, or watching video, is an active experience only for students who have either special training or aptitude. Students mostly are not engaged. Learning remains passive.

The current media available in classes bore students. Many students fall asleep in lectures or while reading. Most of their lives students today spend more time watching television than reading; classrooms are not interesting to many of these students. In the United States, many students do not purchase textbooks for the courses they take, and often do not retain what is learned for long periods after cramming for tests.

2.1.5 Lack of creation of knowledge.

Discovery learning is important for retention. Students learn better if they construct their own knowledge.

Students in classes today seldom create their knowledge. Discovery methods are very difficult in the lecture environment at all levels of learning. Curricula that have depended on discovery, such as the elementary science material developed at Berkeley, have not been successful in typical school environments. Again, the numbers of students, and their need for individualized help, is the barrier. Even science laboratories, often mentioned for their hands-on approach, are mostly cookbook, with little student creation.

2.1.6 Our courses are old

Many courses at all levels have old content. We can see this clearly in textbooks. Thus beginning United States physics texts have changed little in 40 years, except that they have become steadily longer. Although many experts believe that 'less is more' should apply to learning as well as architecture, our learning activities do not show this happening.

Recently Project 2061 of the American Association for the Advancement of Science has conducted several projects examining science and mathematics texts for secondary school. The results are depressing, particularly because so many courses at this level are highly dependent on these texts.

An aspect of old content is the focus on many facts and student memory, rather than on more important intellectual skills. This focus is pointed out in the 2061 studies. Another aspect is the problem of political control that has no place in schools. The poor showing for teaching evolution illustrates this.

2.1.7 Learning systems in the world are mostly fixed-paced.

As mentioned often, our current approaches to learning mostly keep the times fixed, and vary the amount of learning. But this is not a rational approach. It has been imposed for the convenience of teachers and administrators, before computers were commonly available.

People are different and may require different times for learning. We want everyone to learn everything, even if some learn faster and some learn slower. We should hold learning fixed and vary time.

2.1.8 Inadequate assessment is typical.

Current courses assess student learning infrequently and in modes that do not assist student learning, such as multiple choice. Testing also leads to negative student attitudes toward learning. We have known many students that are afraid of exams and do not perform at a level indicating their level of learning. Exams are often used as threats, a poor form of motivation, a stick rather than a carrot.

2.1.9 Few students enjoy learning

Many students today do not like to learn. It is not uncommon to hear students in the United States refer to schools as prisons. University students are often only there to get degrees, and have little interest in learning. This not only negatively influences current learning, but also restricts the possibilities and interest in future learning, important today. People spend as little time as possible in activities they do not like.

2.1.10 We do not have enough good teachers and professors

Our learning systems are based on teachers and buildings, both in schools and in universities. But the supply of teachers is inadequate, and likely to continue to be so.

In the United States we need hundreds of thousands of schoolteachers in the near future, and we are not currently producing enough teachers to meet that demand.

Similar situations exist elsewhere. Further, many people leave teaching for other occupations that are more pleasant and pay more. The buildings are often not there either. The situation is much worse in the developing areas, as already noted.

The same situation exists at the university level. Sir John Daniel, Vice Chancellor of the United Kingdom Open University, has forcefully pointed this out.

> In the last seven days, somewhere in the world, a new university campus should have opened its gates to students. Next week, in a different location, another new university ought to begin operations.
>
> At the end of the millennium in which the idea of the university has blossomed, population growth is outpacing the world's capability to give people access to universities.
>
> There are similar problems at other levels of learning.

Mega-Universities and Knowledge Media –
Technological strategies for Higher Education Kogan Page,
London, UK, 1996.

2.1.11 Learning is too expensive

We have always had examples of excellent learning. But in most cases, even in the wealthy communities, good learning is too expensive for either individuals or societies. Governments talk about improving education, but are reluctant to spend the money needed to improve the current educational institutions. In poor countries very little education is affordable.

> If you do not change your direction,
> you are likely to end up where you are headed.

Ancient Chinese Proverb

2.2 The visions

> Education is . . . the key to . . . development that is both
> sustainable and humane, and to peace founded on
> mutual respect and social judgement. Indeed in a world
> in which creativity and knowledge play an ever greater
> role, the right to education is nothing less than the right
> to participate in the life of the modern world.

Education for all, UNESCO, Amman, Jordan, 1996.

> . . . education is an essential human right, a force for
> social change - and the single most vital element in
> combating poverty, empowering woman, safeguarding
> children from exploitative and hazardous labour and
> sexual exploitation, promoting human rights and
> democrarcy, protecting the environment and controlling
> population growth.   Education is a path toward
> international peace and security.
> 130 million children in the developing world are denied
> this right [education] - - almost two thirds of them girls.
> Nearly 1 billion people are illiterate - - the majority of
> them women.

Kofi Annan
Secretary General of the United Nations
The State of the World's Children 1999

The problems just reviewed and the importance of learning for our
future suggest a set of visions for learning, the other side of the coin to
the problems. These visions should provide us with the direction that
Alice did not have.

2.2.1 Everyone should learn

The first goal for learning proposed is that EVERYONE SHOULD
LEARN to full capacity and desire. It implies that the poor of the
world have learning opportunities as good as the wealthy of the world.
It implies that learning availability should not depend on gender, race,
economic factors, age, previous learning, or any of the other items that

distinguish one human from another. It implies that learning is available in all subject areas, from birth to old age.

This goal also means that learning should include the intellectual skills essential for today, such as creativity, intuition, and problem solving. Learning should include experiencing the great creative products of the human mind, Beethoven's music, Shakespeare's plays, Newton's Laws of Motion, Henry Moore's sculpture, among many others.

It implies that everyone should want to learn and that learning should be enjoyable for all. This is democracy in education, giving equal opportunities to all.

Some of these ideas will be expanded further in the next sections. Note than we emphasize learning over teaching and instruction.

2.2.2 Individualized Learning

Educational psychologists have long pointed out that students are different, in many different ways. They have, at any point in learning in a subject, different backgrounds of previous learning and different interests. They have different learning styles, although we still have much to learn in this area. An effective educational system should treat each student as a unique person. We will discuss this issue again later in more detail.

A learning sequence that is effective for one person may not be effective with another. So in an individualized learning environment providing several approaches to learning a given topic may be essential. We may eventually be able to pick the most effective learning approach for a given individual, but this is beyond our present capabilities.

We can also expect better retention. The learning sequences can occasionally check on previous learning, and give additional help if necessary. So learners will know more after a long time.

2.2.3 Individualized learning time

Another factor in individualization is that students learn at different rates, perhaps differing from subject to subject, and differing with changing interests outside of school or other learning locations. Thus someone might be ready for advanced mathematics at an earlier age than another student. Many students in an individualized learning environment, tailored to the needs of each student, might finish a subject much earlier than is typical today. We expect that most students will progress much faster in an individualized system, shortening the time for learning.

We envision a good future educational system as allowing all students to move at different rates through learning, each progressing as natural for this individual student. This implies that the concept of grade-level will vanish in our future systems. Further, students will probably be at different learning levels in different subjects.

These last two considerations imply that realization of our goals requires a continuing way to evaluate students, almost moment by moment, to know when a different approach is needed, and to know when it is time to proceed to the next learning sequence. This is critical.

2.2.4 All students should succeed in learning.

Success is important in learning. This implies that all learners should master fully each lesson, each subject, even though it may take different times and different learning material. Mastery can be achieved through treating each learner as a unique learner and continuing to work on a subject until mastery is attained, as measured by the evaluation approach just mentioned. Learning activities should focus on learning, not time, as already suggested.

2.2.5 Learning should be enjoyable

It is important that learning be a pleasant activity for all students. Continual success in learning is important in this regard.

A critical factor in learning is the amount of quality time devoted to learning. So we want students to stay at learning, even difficult learning, for long periods of time. A major factor in encouraging this is that students should enjoy learning.

Another factor that should be considered that that we want to encourage future learning, our next topic. If people have enjoyed learning in the past, they are more likely to consider it in the future.

2.2.6 Learning should be lifelong

We live in a rapidly changing world, and people live longer. The median age is rising.

At the moment the median age of the world's
population is about 26 years.   If the trajectory
anticipated by the current United Nations low-variant
projections were to come to pass, this figure would rise
to 44 years.

Nicholas Eberstadt
World Depopulation
The Milken Institute Review, First quarter 2000

A person may have several different jobs and avocations during a
long lifetime.   Schools and universities are not enough.   Learning
needs to be a continuous activity from birth to death for everyone,
perhaps with intervals where learning is at a low level because the
person is engaged in other absorbing and important activities.   One
science fiction novel, *The Troika Incident,* by James Cooke Brown,
mentioned in more detail in chapter 4, portrays a community in which
learning almost stops while people are raising their children, for
example.

2.2.7 Learning should be affordable

The individual and the world must be able to afford learning,
everywhere, for everyone, and at all levels.   A future learning system
must be economically possible.

The critical factor is the cost for an hour of student learning,
considering all costs. It is very important to include every cost
associated with learning in making this calculation of costs.

2.3 Need for a new approach to learning

These goals cannot be attained with our current systems of either
traditional or distance learning.   We need a new form of learning to
reach our goals.

We will explore such a form in this book.   This new form of
learning has great potential for realistically reaching the visions just
described.   But as we will mention frequently we need much more
experience and research with this new proposed approach.

Our present educational institutions are at the service of the teacher's goals. The relational structures we need are those which will enable each man to define himself by learning and by contributing to the learning of others.

Ivan Illich
Deschooling Society
Harper and Row 1971

Chapter 3

FACTORS OF DISTANCE LEARNING

Distance learning, where the learner can be anywhere, anytime, is an important component of the future learning system discussed in this book. Often people have a limited view of the various types of distance learning that can be conceived; for some, distance learning is 'what I do.'

We need to consider the full range of possible types of distance learning. This chapter will provide that examination. Many different factors determine just what kinds of distance learning are to be found, or are possible. Most of these choices are spectra, not just a simple yes/no choice.

These myriad possibilities lead to a variety of distance learning strategies. Some are closely aligned with the visions in the last chapter. The following chapter discusses some realizations of these possibilities of distance learning, and later chapters develop a new possibility.

> Although distance education is a 'hot' topic, how well do we understand it? Our definitions and expectations of distance education tend to be fuzzy. New firms, alternative organizational models, and venture capital funds are emerging with startling rapidly, further complicating matters.

Diana Oblinger and Jill Kidwell
Are We Being Realistic?
EDUCAUSE Review, May/June 2000, page 31-39

3.1 Learning paradigm

A key factor in learning is the underlying view of the learning process, the paradigm. The developer or deliverer of learning units seldom actively makes the choice of learning paradigm. This important choice is almost always an unconsciousness decision, influenced by the current dominant paradigm. Our concern is with paradigms for distance learning.

3.1.1 Information transfer

Almost all learning efforts for centuries, both conventional and distance learning, have followed a paradigm that can be called the information transfer paradigm. It sees the learning activity as transferring information for a knowledgeable individual to a student, through some intervening media. The focus is on information.

In schools, universities, and training, lectures, video, and print material are the dominant media for information transfer, with lectures the most common. In distance learning following this paradigm the dominant media is video, often similar to lectures, or print-based computer units, as in current web courses online. Verification of leaning is based on testing of memory of either information or procedures.

We have argued (EDUCAUSE Review, January/February 2000, and elsewhere), that this paradigm is inadequate for learning in the new century. Learning today has many needs that can not be met by information transfer, including learning problem solving, intuition, and creativity. Further many of the students using information transfer as it now exists do not learn fully. Memory is not enough in the twenty-first century. We need to look beyond information transfer, changing the paradigm for learning.

3.1.2 Tutorial learning

The new paradigm that seems desirable, for distance learning and all learning, is tutorial learning. Human tutors have long been used. Perhaps the most famous example of tutorial learning is 2500 years ago, with Socrates. Oxford and Cambridge used the tutorial method. Every major physicist in 19th century United Kingdom, including Maxwell, Tait, and Kelvin, studied with the same tutor at Cambridge, Hopkins. Many other examples can be sited, mostly for the children of the wealthy who had their own tutors at home.

Tutorial learning with skilled tutors is often effective for learning, but it is very expensive and we do not have nearly enough skilled tutors. We could never have learning with human tutors as a dominant learning paradigm today.

But modern digital and communications technologies now make tutorial learning possible, with the computer as the tutor, for all for distance learning, if suitable highly interactive learning material is developed. This will be a major enterprise.

We will return to this possibility of computer-based tutorial distance learning in more detail in chapter 7 and in other sections of this book. Changing the paradigm will not be easy.

Peter Drucker commented recently, "change is an opportunity, not a threat." (http://pfdf.org/publications/news/may2000/news.html.

> Perhaps all of the stakeholders of higher education are caught in a paradigm paralysis. We have difficulty changing the way we think and the way we believe . . . To break this barrier we should consider asking some new questions.
>
> Michael Hooper
> The Transformation of Higher Education
> In Oblinger, D., S. Rush, The Learning Revolution
> Asker, Boston, MA, 1997

3.2 Level of Interaction

A critical question for learning is how active the student is. Most educational psychologists agree that active learning is better than passive learning. In a lecture environment, particularly in large courses, often very little time is allowed for each student to ask questions. Questions asked by the instructor are often rhetorical, and few students answer them. The activity is called teacher-centered, with the instructor talking almost all of the time. Reading for most students is also a one-way activity. Today learning on the web is primarily with reading or watching video, with little interaction.

Communication in two directions is essential for best learning, but it is seldom found in learning today. It does happen with good human tutors, and it can happen with computer-based tutors. Distance learning might involve various levels of interaction.

3.2.1 Computer based interaction

In tutorial learning with computers the interaction is intense, with both the computer and the learner playing a very active role, and with each interaction being of high quality.

This will be discussed in detail in Chapter 6.

3.2.2 Peer interaction

Peer learning provides another possibility for interaction in distance learning. When a small group of students (perhaps four) work together in a peer learning environment, perhaps with a highly interactive computer program, the interaction is high; it could be called conversational. These groups could be close together physically, or could be remote groups connected through the technology.

A given example of distance learning can be evaluated in terms of the interaction possible for each student. Both frequency of interactions and the quality of each interaction are important in considering the level of interaction as discussed in Chapters 6 and 7. Again we find spectra, wide ranges for each of these factors.

3.3 Student Locations

Distance learning implies that the students learning are not physically at the formal institutions of learning, such as schools, universities, and corporate training centers.

This leaves many possibilities as to the locations of the students.

3.3.1 One or many locations

The remote students may be clustered together, perhaps in a series of classrooms in other institutions. This was the situation in the Chinese TV University when visited several years ago. With this arrangement the students are still together, as they would have been at the offering institution, but elsewhere. There may be only one, or many such distant groups.

At the other extreme, very large numbers of individuals may be learning at many individual locations. They might be in their homes, in public environments such as libraries, museums, and shopping centers, or in many formal institutions. They might be sitting under a tree, or at the seashore. Perhaps two or three students might work together in groups, gaining the advantages of peer learning as just discussed.

3.3.2 Close or far

Often the distance learning locations are close together, perhaps in a near community to the originating school or universities. This was the situation for the Stanford University engineering courses, given at nearby companies through video.

The other extreme is that they can be widely separated geographically. The students might even be worldwide, although there are few examples of this kind so far. We expect this to become much more common in the near future, as more effective distance learning units are developed and moved to many languages and cultures.

3.4 Time Constraints

Distance learning can involve various time constraints. They are discussed in the next two sections.

3.4.1 Beginning time

In formal institutions classes almost always begin at fixed times. This comes from a period before computers when management practices and instructor availability and convenience demanded this.

The other extreme is that a learning segment can begin for each student at any time: on any day, twenty-four hours of the day. This is possible with today's technology, but still not common.

3.4.2 Pacing

Formal institutions differ from the possibilities in distance learning, with regard to student pacing. In traditional activities the learning time is fixed for a particular learning segments.

Distance learning, in some forms, allows the pacing to depend on each student. Each student takes a different time to complete each segment, depending on the background and needs of that student.

3.5 Size of Learning Segments

In our schools and universities, the usual length of a learning segment is a course, lasting a quarter, a semester, or a year. The learning material included may depend on the schedule of the institution – the length of the year, quarter or semester. In universities, professors argue seemingly endlessly about the advantage of quarters and semesters.

But another possibility is that segments can be of any length. Distance learning can follow either strategy, although fixed lengths are most common so far.

For flexibility we might find a future situation in which units are relatively small, so students could have different paths in learning, depending on their needs. Students need not be aware of the size of distance learning units. The process can appear continuous to the student, with learning in a given subject lasting for years or for life. In this situation, the concept of 'course' is no longer useful.

3.6 Student control of content

Learning can provide no or limited content choice to the student, as in most existing courses, or it can turn over all choice of content to the individual student. This last possibility is not likely, although some people seem to think it is ideal. Partial student control of content seems more likely.

As an example of an intermediate strategy, the online physics course we developed about twenty five years ago at the University of California, Irvine was based on two sets of material, different approaches to engineering physics. We specified six paths through the material, with the student choosing the path at each diverging point. For a meaningful choice, the student must understand the alternatives, a nontrivial responsibility of course developers.

3.7 years of Learning

We can no longer assume that learning occurs only for the years 5 to 25, the school and university years. In our rapidly changing world learning can and should be lifelong, a continuous activity from cradle to grave, as we emphasize in this book. This should include both very early learning and many stages of adult learning. However, breaks in learning are possible.

3.8 Student support

Distance learning can provide various aids to students.

3.8.1 Helping students in trouble

Students may not have a smooth learning path. They may, for a wide variety of reasons, encounter learning problems. A critical step for efficient learning is to locate student problems, often not understood by the student, and offer appropriate assistance.

A learning system can provide many types of help to the student for learning problems. In older learning this was often provided partially by office hours and discussion sections, not enough for many students. More recently this help has come through electronic means, such as email or chat rooms, tactics that work only for small numbers of students. In the United Kingdom Open University tutors throughout the British Isles provide this assistance.

With distance learning the electronic means are also possible. But another possibility for such help is from the learning material itself. It can actively seek out the student problems, and offer interactive aid. This approach for assisting students will be explored in detail in this book.

3.8.2 Peer learning

An important kind of student support comes from other students, already mentioned. Peer learning, between students at the same point in learning activities, or as tutors to other students, is a valuable aid to learning. In traditional institutions this is seldom openly encouraged.

A distance learning activity may or may not encourage peer learning. Many opportunities are available when computers play a major role in the learning activities. Programs can encourage and aid in the creating of learning circles of students, both physically with students close together and electronically for those far apart. This could be through email, of could involve more detailed interaction.

If the computer is storing frequent records of each student, as we will suggest, background programs can search these records. The computer can encourage students, at the same point in learning, perhaps those having the same learning problems or the same zone of proximal development (considered later), to work together. Or the computer learning programs can encourage students who have learned some material to aid others who are still trying to master the unit, a valuable learning experience for both.

3.9 Teachers or No Teachers

It is often assumed that teachers are essential for learning. But it is clear from everyday experiences and from many studies that much learning occurs without teachers. In childhood marvelous early learning occurs, such as the learning of an initial language. Much adult learning is without teachers.

Distance learning may or may not involve teachers. As the number of students increases to the thousands or beyond, as discussed in the next section, the concept of 'teacher' becomes vague, even impossible. In the system described in this book, teachers in the usual are very unlikely except in special situations.

3.10 Number of Students

A recent online discussion asked about the 'ideal' numbers of students in distance learning environments. Several people on the list, from United States universities, recommended 20 or 25 students. They were basing this on the type of distance learning now common in United States universities.

On the other hand, some of the foundation courses in the United Kingdom Open University have over 10,000 students. Clearly as suggested many forms of distance learning are possible. So a wide range of numbers is possible, with different types of distance learning.

The problems of education are universal, in a world with six billion people, many having no or limited schools available. Distance learning, in some form, may be the best possibility for reaching everyone, at all ages. So we would need to address these problems a form of distance learning suitable for very large numbers. This will be discussed further later, as the system suggested here is for many students.

3.11 Grading or Mastery

In schools and universities almost all students are given a grade. Several grading systems are available, with students spread over a range from good to bad. These grades are determined primarily by examinations, although other methods such as portfolios have also been tried. In large courses tests are, unfortunately, often multiple choice exams (called multiple guess by students.) These grades are taken as an indication that some students have not learned the material. Many distance learning environments are based on this grading strategy. But such cases as Albert Einstein, a poor student in school with low grades, lead us to question the value of grades.

Benjamin Bloom and others suggest another approach. He noted, in extensive experiments he and his students conducted in the Chicago public school, that with a tutorial approach, with human tutors, almost all students could learn. This is mastery learning; almost all students learn everything. Everyone succeeds. The problem he raised in the "Two Sigma" paper was how to accomplish this in a way that is feasible economically for the large numbers of students we now have. We hope to address that problem.

A major difference between grades and mastery is the role of examinations, student evaluations. As noted, they are typically used to assign grades in approaches that require grades. However, in mastery courses evaluation determines what new learning material to present next for each student.

For mastery we locate student learning problems and offer effective assistance offered, perhaps with several different sets of learning material. This is further discussed in Chapter 9.

3.12 Cost Factors

We cannot ignore the costs associated with learning. These must be affordable by the individual and by the society. For learning both development and delivery cost, and other costs, are important. We should better understand the costs of traditional learning and distance learning.

Courses in traditional institutes often cost very little to develop. Most of this cost is in the time the instructor devotes to the course, both before and during the course. The major consideration before delivering the course is often the selection of a textbook.

When these same individuals work in distance learning, the costs are similar, because they are still using the same paradigm. But other approaches might involve very different costs.

With learning, including distance learning, several factors concerning costs are important. Different systems will have different costs. Chapter 12 is devoted to the cost problem, focusing on the costs of the tutorial system recommended in this book.

3.12.1 Cost of Development

A learning segment to be delivered at a distance involves some costs of development. As with the other factors discussed, there can be a wide range of expenses. Often, even commercially, there is not a clear view of total costs. Our experiences are that textbook publishers often not have a complete view of the expenses involved.

We had major curriculum developments in the United States in the period immediately following the USSR Sputnik, over a wide range of subjects and levels. A more recent data point with regard to distance learning is the United Kingdom Open University. So they give us a basis for comparison, used in Chapter 12.

Spending large sums on development will not guarantee high quality learning. However, good units for distance learning will be costly.

3.12.2 Cost for a student hour

Costs of development are not the most critical costs in considering distance learning, although they are the most frequently mentioned item. From the standpoint of both the individual and the society (the country or the world) the most important factor is the cost for a student hour, including development, delivery, possible profit, and administrative costs. In calculating the cost for a student hour of learning it is important to consider all costs, even those sometimes hidden.

Expensive development can lead to high quality learning material, at low costs per hour, if the delivery system is inexpensive and if large numbers of students use the learning units. When we think of the learning problems of the world, and solutions to these problems, large-scale distance learning with many students is likely to be very important.

3.13 Evaluation

Learning material should be evaluated for learning effectiveness, but often it is not. Evaluation can occur at various levels, both formative and summative evaluation

3.14 Delivery Method

Many delivery methods for distance learning are possible, even within the same product, including mail, email, cable, CD-ROM, DVD-ROM, Internet, local wireless delivery, and satellite. A given product may be delivered in several ways.

The nature of the learning, and the costs, may determine delivery procedure is to be used. We will discuss the delivery methods available today and how they could be used. For flexibility, we might use several delivery methods for the same material, so we need not pick just one.

3.15 Learning Subjects

A given learning program might offer a single learning sequence, or an unrelated collection of such sequences. It might involve one or several degree programs, or a very wide range of material.

The facility might offer a typical school or university range of units, might be concerned with pre-school activities, after school materials, adult learning, training, new subjects, or some combination of these.

3.16 Credit or NonCredit

Learning may be in an environment that offers credit or certification. Or the learning may be undertaken with no idea of attaining such status. Adult learning is often without credit.

3.17 Profit or NonProfit

The learning segments may be offered for profit or may not be part of a nonprofit organization, such as at a state university.

Chapter 4

DISTANCE LEARNING – PRESENT AND FUTURE

Many types of distance learning are possible, as discussed in Chapter 3. These factors can be put together in many ways to produce a distance learning system. We review a few of the widely used existing possibilities in this section, and then consider some interesting possibilities found in science fiction. Later chapters will present the approach to distance learning recommended in this book.

4.1 Correspondence Learning

The use of ordinary mail is one of the oldest ways to distribute learning units, in print, for distance learning. Other materials such as CD-ROM can also be mailed. Examinations can be distributed in this fashion. This was a classic approach to distance learning. Components of this approach are used in many other approaches.

4.2 Remote Sound and Video

The use of radio and video to deliver distance learning has a long history. Many years ago in the United States there was a program called Sunrise Semester. Major projects in Africa by the World Bank and others used battery-driven radios.

The most extensive television learning activity, and perhaps the largest university in the world, was and is the Chinese TV University, now with over a million students. When I saw this activity about a dozen years ago the cameras recorded live lectures, talking blackboards, often only a little ahead of their broadcast to students. The students were in classroom environments all over China, often in company education centers, with the TV set replacing an instructor at the front of the room. Occasionally someone came to answer student questions. I do not know if this is still the procedure followed; it seems likely to have changed with current progress in China. In addition to national production, some of the television programs are produced in regional centers such as Sichuan.

A major effort in the United States to produce distance learning units with video was the Annenburg-CPB (Corporation for Public Broadcasting) project, which produced many video-based courses. These courses had various auxiliary materials, mostly print. Their production was well financed and professionally done, so the video quality was high. The programs were and are often shown on the Public Television Network, independent of formal courses. We suspect that this use is the largest use. Colleges were free to offer these materials as courses, perhaps providing some student aid.

In the United States one of the longest and best known efforts of this kind in a specialized area is the Stanford University engineering program, bringing courses to groups in technical companies in the San Francisco area, via video. Mostly instructors were photographed while giving lectures. The students in this situation were highly motivated, as the material was useful to their future success. They were already experienced students. They could ask questions via a phone connection to Stanford. This approach has been expanded to other areas.

Initially Jones University proceeded with video delivered via the television cable, owned by Jones. Videotape and CD-ROM allowed a more flexible approach, as the student could stop the tape and review a section of the material. Some programs used two-way video, relatively expensive and still unproved as to value in learning.

In all these situations, the video was very similar to a lecture, perhaps identical. So the dominant paradigm was information transfer. In almost all, assistance and feedback to the individual student was limited.

Proponents of this approach usually see current Internet bandwidth as the major problem, as they want to see video delivered on demand at full video rates. They promote higher bandwidth.

4.3 Remote Versions of Standard Courses

Another scenario for distance learning growing rapidly particularly in universities in United States and Europe at present is to imitate current courses in universities on the Internet, through the World Wide Web. The same approach can be used at pre-university levels.

Online learning or e-learning are common names. The vehicle for delivery is usually a Web browser, Internet Explorer or Netscape, usually displayed with all the usual bells and whistles, icons and pull-down menus, even when these have no relevance to the learning activities.

The typical beginning is an existing course taught in a conventional fashion in a conventional institution. The instructor may start by placing the course syllabus and related information on a Web site. Usually a textbook is involved, and homework may be submitted electronically. If the faculty member involved is writing a textbook, this may be online. A book written by someone else may be similarly available. If bandwidth permits, streaming video of lectures may be part of the units.

Mail and CD-ROM could distribute lectures and print. A part time programmer may assist the faculty members in constructing Web pages.

Developmental costs appear to be low, partially because the faculty is not paid in addition to regular salaries for their developmental efforts. The faculty costs are typically ignored in considering the cost of development. The cost per student hour is high, because the faculty member 'delivers' the course.

The learning paradigm is almost always information transfer. There is an instructor, just as with standard courses, in control. E-mail, list servers and newsgroups are a common way to provide assistance to students. Chat rooms and listservs may be available. When funding is large, two-way video conferencing may be used.

Groups of twenty or thirty students are common with this approach. This type of distance learning will have great difficulty scaling to large numbers of students, because of the need for an instructor, because of the difficulty of supplying individualized attention, and because costs per student may rise as more students are involved.

In many places university administrators encourage, even demand such efforts in producing online courses. They apparently see this as a cheap way to offer distance learning courses, increasing student outreach, although they do not appear to have done a careful financial evaluation. Such a study might suggest that this is not a cheap way to proceed in the long range. Although the costs for individual courses are low, the total investment in this direction is considerable. The administrators in universities should look carefully at this expense.

A number of organizations have formed to provide lists of such sites. Some, such as the California Virtual University, have already come and gone. Others, such as the Western Governors University has had limited success so far. A number of organizations are supplying related software, mostly for management, and will host the web sites.

Another question to raise is whether universities have the necessary skills for this endeavor. Bob Heterick, in The Learning MarketSpace of April 2000 comments that "there is nothing to suggest that institutions of higher learning posses the skill set, management flexibility, or entrepreneurial reward structure to be successful in the creation and maintenance of customer-attractive, asynchronous learning   products."(Available   at http://www.center.rpi.edu/Lforum/LdfLM.html.) Faculty have little experience in developing curricula.

4.4 The United Kingdom Open University

A very successful example of distance learning today is the United Kingdom Open University. It has a headquarters facility at Milton Keynes, about fifty miles north of London. It has influenced many other institutions. The Open University, starting in 1969, now has alliances and centers in the United States, at Florida State University and with the Western Governor's University. It has many students outside of the United Kingdom.

The Open University offers a full range of degrees, and many adult education programs not leading to degrees. It has over 200,000 students, many seeking degrees, in a country much smaller than the United States. It is open to all students, without prerequisites.

An important clue to understand the success of the Open University is the way they prepare learning units. They have a careful full-scale system for producing their courses, developed over many years, involving a team of developers.   Several years and millions of pounds go into developing each course.

Evaluations of the courses are an important part of this developmental process. Courses are improved before use. Their effectiveness is demonstrated. A course is redone in seven years. In rapidly changing areas updates may be sooner. Few distance learning institutions consider this important aspect.

A key aspect of the Open University delivery of courses is the tutor, available in centers around the country. They represent the main possibility of individualized feedback for the students. Sometimes these tutors are from the traditional universities. This is an expensive component of the delivery system. It is difficult to extend worldwide. More students imply more tutors, so greater costs.

> The tutorial support system is a key element in the UKOU's success. . . . this system is difficult and costly to reproduce in other countries where the UKOU might like to operate.
> The UKOU is eager to discover . . . whether the knowledge media can speed course production and provide tutorial support that is less geographically based.
>
> Sir John Daniel, Vice Chancellor of the Open University
> Megauniversities and Knowledge Media
> Technological Strategies for Higher Education
> Kogan Page, London, UK, 1997.

Initially Open University courses were based primarily on print material and television, the major media available at the time of formation. The British Broadcasting Company produced the television material, but later the Open University had its own studios. The video was superior to most education video because of its professional production standards.

In the early days the video material was broadcast on BBC, but later video material allowed students to escape from the fixed schedules of broadcast facilities. Testing showed that the video was not very effective in assisting learning as had been hoped, so it was less used in later courses. In some courses, such as the recent geology course, it is pedagogically important, and distributed as part of the course.

Recent courses, such as the foundation course in computer science, use more technology. But we do not believe that the Open University has developed a full computer-based tutorial course. The human tutors spread over the country remain the model for providing detailed assistance, a strategy that appears difficult for them to change.

In spite of the high costs for developing each course the cost per student is much less than that of traditional universities, an important ingredient in the political strength of the institution. It has survived unfriendly governments. We return to these considerations of cost in Chapter 13. The Open University graduates are competitive with those from traditional universities.

The success of the Open University is clear, and it is a major step in distance learning institutions. It shows that large numbers are possible, economically, in distance learning.

The reader should consider the striking differences between these last two approaches. Distance learning, as suggested in the previous chapter, can mean many different things. These two do not exhaust the possibilities.

4.5 The Mega-universities

The Open University describes other universities called mega-universities that partially follow the directions of the Open University. A mega-university is defined to have over 100,000 students, and to use distance learning. Interestingly, although these institutions are scattered over the world, none is in North or South America.

The largest mega-university is the China TV University System, already mentioned in this chapter.

The Centre National d'Enseignement a Distance (in France, over 50 years old) has courses ranging from primary to postgraduate. Courses are primarily correspondence courses. Telephone feedback is important in recent efforts. Also in Europe is the Universidad Nacional de Educacion a Distancia, in Spain. Print and telephone are again important.

We will only list the additional mega-universities. Further details are in Sir John Daniel's book, *Mega-Universities and Knowledge Media – Technological Strategies for Higher Education*, and in the references available there.

- The Indira Gandhi National Open University – India
- Universitas Terbuka – Indonesia
- Pa.yame Noor University – Iran
- The Korean National Open University
- University of South Africa
- Sukhothai Thammathirat Open University – Thailand
- Anadolu University - Turkey

All the examples of distance learning presented so far are real. These existing programs do not use all possibilities or fully use the technology available to us. But we also have much to learn in looking at imaginative descriptions of distance learning. This is our next activity.

4.6 Distance learning in fiction

This section and the next consider three visions of the future of learning from science fiction and one from fictional chapters in a nonfiction book. They are not complete accounts of a new learning system, and differ greatly in the details presented. They also project forward different periods in the future. We discuss them in the order of their writing. We believe these views have much to tell us. We will see several learners, Sally and Nell. After presenting the four sources, we look at common features.

4.6.1 Arthur Clarke – The City and the Stars

This early novel has little about learning. However, the sagas, a well-developed virtual learning environment, are a very interesting learning environment. They are the ultimate in virtual reality, perhaps the origin of the StarTrek holodeck. Groups participate jointly in adventures that often have strong learning components, through the central computer. Many sagas are available. They show a possible future form for experiential learning.

4.6.2 George Leonard – Education and Ecstasy

This book, from 1968 (just before the formation of the UK Open University) has two fictional chapters about a school of the future, showing a visit of parents to their children's school.   Leonard presented a differing but related fictional view of learning in a 1984 article in Esquire. Both are in the 1987 edition of *Education and Ecstasy*.

An interesting chapter that precedes the fictional chapters raises the question of why there should be a school. Leonard says, "Practically everything that is presently being accomplished in the schools can be accomplished more efficiently and with less pain in the average child's home and neighborhood playground." Leonard means that we do not need current schools. He advises parents to visit their children's schools, on an ordinary day, to see what happens.   However, he believes we can do better, leading to the account of a school in 2001. Unlike the other visions in this section, the school still exists.

The school in *Education and Ecstasy* separates knowledge aspects from affective aspects, with different approaches. The school is very different from today's schools. "While the children are on the school grounds, they are absolutely free to go and do anything they wish that does not hurt someone else. They are free learners." There are no required scheduled activities. Learning happens because children enjoy it.

Knowledge learning occurs in the basics dome, one place where students may choose to go. There are no classrooms or teachers. Large computer displays are at the walls, recognizing the student. Stored records of students' learning histories show where learning is to begin, and what the student problems are. Both keyboard and voice input are used. Ongoing Brain-wave Analysis finds if the student understands or not at each point, so student progress is rapid.

The student-computer interaction in learning, called computer-assisted dialogue, is tutorial, as the situation closely resembles the interaction between a student and a skilled human tutor. It is highly interactive and individualized, as all learning should be.

Students move at individualized paces through the learning experiences. The amount of time allocated in a single session at the computers for the student depends on the student background and on the number of waiting students. There is limited human interaction, as the learning programs bring neighboring students together occasionally.

The parents find their daughter, Sally, three years old, working on language. The four-year-old girl on one side is "dialoguing about primitive culture," and the six-year-old on the other is learning elementary calculus. Note the ages and subjects!

We see only a little of Sally's interaction with the computer. Here is the beginning: "I hear Sally saying 'Cat' into her microphone. Almost instantly, a huge grinning cat's face gathers form and the word "cat' appears at the bottom of the display." In the following dialog, Sally is queried about linguistically possible alternate spellings of 'cat.' After a bit, Sally goes to find some friends, and plays a game. Remember that she is a free learner, although she is in school, so she can do whatever she wants to do.

I will not review the affective part of the school. It has some interesting aspects. The hands on way of learning science, literature, and history offer useful suggestions for learning.

The school is for children from three to ten. The content is beyond that of a ten-year-old today, as we learn from the student learning calculus. Leonard says "after the age of ten they may never again attend a separate, formal, degree-granting educational institution. This is not to say that they will stop learning. On the contrary, they will be free to begin a lifetime of learning in a society dedicated to education." School is a preparation for lifelong learning.

The school described In the Esquire article puts greater emphasis on planning student's time and activity. The morning begins with this planning. Students can work partly at home.

4.6.3 James Cooke Brown - The Troika Incident

Unexpected time travel is the mechanism in this 1968 book for going to about 2070. The world is not homogeneous, so learning activities differ in each community.

We see Loma Verde, about four hundred people on the California coast. "The real business of the people of Loma Verde is not raising artichokes, or grapes, or even feeding themselves. It is raising children . . .. there is no formal school, no teachers, no grades to pass, no marks. The Companions (citizens of Loma Verde) do their own teaching in small study groups, or in tutorial relations with individual children . . . But mostly the children simply range for themselves over the cultural landscape the adults' activities create. . . . in the free but stimulating atmosphere there Loma Verdians seem to have created for their children, they all love to learn . . ."

". . . the key to this autonomous learning . . . [is that] Their children teach themselves how to read . . . when they are between two and four years old. . . the language in which they learn is Panlan . . . spelled phonetically . . . [and] very simple grammatically. . . Once a child has learned to read in this simple international tongue, books, ideas, the regional language and its literature, all seem to come pouring in"

> . . . reading never loses its each grace for them. And the generalization of this early attitude – that all learning will be joy – seems to infuse their lives.

The community has a Children's House with "private studies, labs, and workshops for them all." It is there that they teach themselves to read.

> " The two year olds were racing back and forth between two large toys that looked a little like typewriters. Only no keys - - - as they shouted in Panlan at these toys their words would immediately appear all neatly typed out on a large screen - -   They called this writing - - - they could either write these words for themselves - - - or they could send messages to each other"

Reading and writing are critical in this society. It happens with a computer, beginning at two.

A common device in this future society is the reader, " a light flat, plastic box with a glass screen . . . [with] an alphanumeric keyboard." The search mechanism is familiar to today's database user. "Every book that has ever been written . . . is waiting to materialize in this little box sitting on you lap." The description sounds like the Internet possibilities. But it goes beyond today's Internet, with no commercial component. "you can learn anything, find anything out, look up anything, simply by fiddling with a little plastic box sitting in your lap." A storage area in the Australian desert holds all the information. We can assume that this device plays a major role in adult learning.

There are still universities in this society, although there is nothing we would recognize as a school. We do not hear much about them. Generally people do not attend them until they are finished raising their children, as noted.

4.6.4 Neal Stephenson - The Diamond Age

This is the most recent of the four sources considered, published in 1995. Learning (150 years from now?) is central to the novel, so we see much more of it in this book than in the others.

A wealthy and powerful man is worried about the education of his granddaughter, Elizabeth. He talks to an excellent engineer about the problem.

> . . . to raise a generation of children who can reach their full potential, we must find a way to make their lives interesting.   . Do you think that our schools accomplish this? Or are they like the schools that Wordsworth complained of?

His answer is that schools are not adequate. With support of the powerful person, the engineer develops a 'book,' The Young Ladies Illustrated Primer. Two copies of this book are produced initially, one illegally, and one falls into the hands of a poor four year old, Nell, living in unhappy circumstances. Another is later produced for the engineer's daughter, Fiona. Later, hundreds of thousands are produced, all for girls.

The first time Nell opens the book it begins with a story about Elizabeth, the granddaughter of the wealthy person. The book quickly learns that this is Nell. The engineer says, "It is unlikely to do anything interesting just now. It won't really activate itself until it bonds . . . it's looking for a small female. As soon as a little girl picks it up and opens the front cover for the first time, it will imprint that child's face and voice into its memory." The device has a camera.

The book bonds with Nell. The first learning session is portrayed in a marvelous passage.

> The book spoke in a lovely contralto . . . Once upon a time there was a little princess named Nell, who was imprisoned in a tall dark castle on an island in the middle of a great sea, with a little boy named Harv, who was her friend and protector. She also had four special friends named Dinosaur, Duck, Peter Rabbit and Purple.
>
> Princess Nell and Harv could not leave the castle. But from time to time a raven would come to visit them.
>
> 'What's a raven,' Nell says? The illustration was a colorful painting of the island seen from up in the sky. The island rotated downward and out of the picture, becoming a view toward the ocean horizon. In the middle was a black dot, and it turned out to be a bird.
>
> Big letters appeared beneath. 'R A V E N,' the book said. 'Raven.' Now say it with me.
>
> 'Raven.'
>
> Very good! Nell, you are a clever girl, and you have much talent with words. Can you spell raven?
>
> Nell hesitated. . . . After a few seconds, the first of the letters began to blink
>
> The letter grew larger until it pushed all the other letters and pictures off the edges of paper. The loop on the top shrank and became a head, while the lines sticking out the bottom began to scissor. 'R is for Run,' the book said. The picture kept changing until it was a picture of Nell. Then something fuzzy and red appeared beneath her feet. 'Nell Runs on the Red Rug,' the book said, and as it spoke, new words appeared.

This passage continues in this interactive way. The reader can find the original beginning on page 94 in the paperback edition.

The first time Nell opens the book, it already knows quite a bit about her! Her brother Harv and her toys occur immediately, within a fanciful story, told to her with pictures in the book. Her mother and her mother's horrible boyfriends also are in the legend the book presents.

It was not know that Nell would be the user. Nell has not used the book before, but it has been listening to her and to what has been happening around her even before she used the book for the first time! It even took a picture of her. It has begun storing a record of her situation and interests, soon augmented with information about Nell as a learner, obtained as she uses the book.

The basic story of Princess Nell (or another person) is in the program. At Nell's next session, there is a summary of the entire fantasy, including a happy conclusion. So Nell knows from the beginning how the story will end. The book always talks about Princess Nell.

As with *Education and Ecstasy* and *The Troika Incident* learning to read is the basis of all further learning, and again it occurs at a very young age. The book remembers what progress Nell makes, and what her learning problems are. When the story speaks of a raven, Nell, living in a slum, ASKS what a raven is. The book hears her, and shows her, with a picture. It quickly determines that she cannot read, and begins to help with the alphabet, highly interactively.

The book has other skills. It can record and illustrate things that Nell tells it (writing.) It teaches Nell the art of self-defense, and helps Harv and Nell escape when their situation becomes life threatening. It can defend itself against those who try to steal it.

Nell learns to cook healthy foods. The book contains an encyclopaedia. It displays books, like the machine in *The Troika Incident*.

It has fantasy situations that encourage problem solving. Nell participates in these situations, as Princess Nell, as with Clarke's sagas. "As she climbed the switchbacks, she forded those delightful current of air over and over . . . the little shrubs that clutched rock and cowered in crack became bigger and more numerous . . . 'Nell looked for a safe way down,' Nell essayed. . . . 'No, wait!' she said.'" These situations become more important as Nell approaches the end of her education. They even include learning to program 'Turing machines.'

The book (a computer without keyboard) has paid outside help for voice output, humans who work with the students. We see Miranda whose lovely contralto voice Nell heard when she first opened the book, and continues to hear. Miranda is a ractor, a technology-based actor. Nell has no contact with Miranda; the interaction is through the computer, from a script. They meet at the end of the book as the fantasy of Princess Nell and Nell's own story merge.

The engineer regards this need for a human voice for speech output as a design problem. He is unsatisfied with the quality of computer generated voice. When he must generate hundreds of thousands of copies of the book, and so ractors are not practical, he uses computer-generated voice. These users of the book become Nell's 'mouse army.'

Nell and the other two ladies with the book do go to a fancy school, Miss Matheson's Academy, one that would not normally be available to someone of Nell's background. They are the three brightest students and continue to work with the computer. Nell is the best, perhaps because of the real-life problems she encountered because of her poor environment. The wealthy person attributes this difference to the ractors. In Elizabeth's case, many ractors were involved. For Fiona, it was mostly her father, the engineer.

No boys learn with the book. Different programming would be needed, since the fundamental story is perhaps oriented toward girls. The possibility that this could be the educational system for all is not considered, except by implication in the comments by the wealthy person. However, this is a novel, not a treatise on learning, so it is unfair to complain about such things

4.7 Aspects of these Fictional Views of Learning

We see interesting features in these fictional accounts of learning that can offer guidance for the future. None presents a complete system for learning, but they give us many ideas for new approaches. The reader can compare these with the visions in Chapter 2, and with the form of distance learning proposed in this book.

4.7.1 Love of Learning

These accounts stress that learning should be enjoyable, that everyone should like to learn. This encourages lifelong learning, essential for our changing world.

Perhaps one reason is the flexible schedules. Children are not required to do anything. They engage in learning activities when they want to. They do not sit for long periods in assigned seats, not talking. Learning activities are very enjoyable.

4.7.2 Learning to Read and Write at an Early Age

Learning to read and write occurs at a very early age, between two and four years old, and with NO human teachers! All students learn to read. No one is illiterate. Learning to read closely follows learning to speak, a natural progression in language learning. Reading is critical to later learning, with love of learning.

Writing comes next in several of these accounts. With Brown the two happen together. Writing is done by talking to the computer.

The computer may eventually eliminate the need to read, because the computer can do all the reading. But this is not the present situation.

4.7.3 Tutorial Learning

Much of the learning is tutorial; the interactions with the student are close to that between a tutor and a skilled human tutor, with frequent interaction in both directions.

Like human-tutor interactions, these interactions use our native languages. The examples are in English or Panlan, but other languages are possible. Pointing, a limited form of interaction is little used, in contrast to modern Web learning material. The student replies to questions, or asks questions, with no restrictions on the language used. Input is free form.

4.7.4 Computers as Critical in Learning

All these educational systems are computer-based. The computer is the tutor. They examples we see go beyond current use of the computer in learning. However, our current technology is adequate for much of what we have seen. The problem in attaining such learning systems is not the technology, but the development of the learning material and the social and political structures required.

4.7.5 Communication with Voice

Communication with the computer is through voice. People talk to the computers, and the computer talks to them. Voice input is accepted as a standard procedure, but with Stephenson the quality of computer-generated voice output is questioned; the voice Nell hears is that of Miranda, but computer voice is sufficient for the 'mouse army.'

4.7.6 Adaptive learning

These learning experiences continuously adapt to the needs of each individual learner, offering a unique learning path. This is in contrast to what happens today in schools, universities, and other formal learning environments. We see this with Sally in Leonard and Nell in Stephenson, although we do not see large numbers of students. When the computer determines that Nell cannot read, it proceeds accordingly. Records are stored frequently and the computer program uses these records to adapt to student needs.

Different learning approaches may be best for different students. One aspect of this is that learning time varies from learner to learner. Another is that each individual embarks on a new topic at a time that is different from that of other learners.

4.7.7 Accelerated Learning

In some of these examples, learning occurs much faster than today. One of Leonard's students is learning calculus. This is a consequence of adaptive learning. Learning is tailored to the needs of the student, so progress is rapid. No students spend time 'learning' what they already know.

4.7.8 Mastery Learning

We all master our native languages, in our first few years. There some evidence that the learning systems in the four books all insist on mastery. Grades do not exist in these environments. Learning and assessment are woven into a seamless product.

4.7.9 No Schools

Current views sometimes assume that schools and universities will last forever and forever. But in these examples schools are not prominent, although some still exist. Even when we see a 'school,' as in Leonard, it is very different than a current school, and he explicitly raised the possibility of no schools. Having people gather at a central location for such knowledge is no longer necessary.

We see little of university education, except to hear that it exists in the society Brown describes.

4.7.10 Distance Learning

If we do not have schools, where does learning take place? Nell learns in her home. In Brown, learning to read and write takes place in a special building for children, not a school. This is distance learning.

The computer makes distance learning possible in these examples. Everything can be made available in homes, libraries, shopping centers, children's buildings, and other informal environments.

Chapter 5

INDIVIDUALIZATION AND INTERACTION

5.1 Individualization

People are both similar and different. We have no trouble recognizing someone unfamiliar as another human being. The genetic background of all people is similar; we have 48 chromosomes, being mapped now by the human genome project.

On the other hand, we can easily recognize individuals, distinguishing them from others, except possibly in the case of identical twins. Many features distinguish each person, eye color and shape, size of various parts of the body, hair color, and numerous others. Our genes are different.

In learning, too, people are both similar and different, both important in thinking about learning. Both are important in learning. We all have similar senses and similar brains, in overall details, so we do not each learn in entirely different ways.

Even casual observation establishes that we are not identical in learning. Various schools of educational psychology emphasize these learning differences. Not everyone learns the same way. Every student is different, with different backgrounds, different learning problems, different interests and different learning styles. There is no guarantee that learning material that works well for one person will work with another. Both the similarities and differences in learning are important in developing learning materials.

Individualization of learning is an essential factor in making learning more effective, allowing us to react to differences in learning. Each student may need different learning materials and different times for full learning.

Another broadly relevant outcome is a growing recognition of, and respect for, the inherent individuality in the structure of human intellect

Roger Sperry, Noble Prize Speech,
December 8, 1981

As suggested, several different learning approaches may be needed, as students have different capabilities.   As we learn more about learning styles, it may be possible to predict the approach best for each student, leading to better learning in less time, but this is not possible at present.   We can try different possibilities. Multimedia approaches will help with many students.

However, if we are to educate all of humanity, we can rely on the idea that these differences are not extremely great.   We do not need a million entirely different learning approaches for a million students. The job of individualizing learning is possible.

To achieve individualization in learning, we need learning material that interacts with each student, finding the learning problems and using the background of the student.   The next section of this chapter discusses how interaction in learning can individualize learning with the appropriate learning material.

5.2 Interaction

The word interaction is used frequently today, particularly with computers. Many computer articles and ads refer to the computer as interactive.   But some of this use does not refer to interactive tactics useful in learning.

Many existing materials on computers, some for learning, are "very slightly interactive". The level of interaction is low, in the sense already discussed.   We discuss highly interactive tutorial software in this book.

Interaction is not a single idea, but a gamut of many possibilities, just as we saw for the numerous forms of distance learning.   We will discuss these factors for interaction in this chapter.

5.2.1 A Paradigm for Interaction

We need a model, a paradigm, for highly interactive learning for the computer. We discussed paradigms for learning in Chapter 3.   The old paradigm for learning is information transfer.   The paradigm we suggest for future learning is tutorial learning.

Our model for interaction is interaction between a skilled teacher and a small group of students, perhaps as many as four. A good tutor, as already mentioned, with a few students can provide an excellent interactive learning environment for the students in this small group.

Socrates, as described by Plato, is one excellent example of a highly interactive student-teacher interaction. He proceeded by asking questions to a small group of students. Student answers led to additional questions. Socrates did not give lectures or write textbooks!

However, as we have had to educate more and more people, it has been impossible to provide active human tutorial learning to all students. We have retreated to less interactive educational modes, lectures, video, and books. These approaches provide spectator education for most students.

Interactive technology offers new economically reasonable possibilities, even with very many students. But these possibilities are largely unrealized. This book considers them.

The type of interaction proposed for the learning style suggested is 'conversational interaction'. It resembles, as seen in our model, the learning conversation between one student or a small number of students and a skilled teacher. Software using this paradigm is 'highly interactive'. Technology can create engaging highly interactive environments for *all* students.

Interaction is a complex issue in learning. We need to consider both the frequency of interactions and the quality of each individual interaction. We discuss both in the next two sections.

5.2.2 Frequency of interaction

The frequency of interaction between computer and student is our first concern, one of two factors that determine the degree of interaction. With the computer and associated technology students, with well-prepared materials, can make meaningful responses every few seconds. The computer program, if prepared with this in mind, can consider these responses carefully, also considering the stored student information to be discussed later.

With such frequent interaction and analysis, the student experience can be highly interactive, almost conversational. We refer to such programs as dialogs or conversations.

Students actively involved in learning do better than passive students. The learning environments created in interactive modules will stress involving students, getting them to do something useful for learning frequently. We want students to be eager active participants, not spectators, in the learning process. We want them to enjoy learning. Retention and learning time are improved.

The student in tutorial leaning is highly active. The learning process is the antithesis of what happens in the lecture or on almost all current web sites.

Interactions must be frequent in both directions, from computer to student and from student to the computer, to locate learning difficulties and to maintain student interest. We find in testing in public libraries, where students are free to leave at any time, that a gap of more than twenty seconds between high-quality student input begins to lose students in learning units. Frequent interaction maintains student interest, even with difficult learning activities, just as it does with a human tutor.

5.2.3 Quality of each interaction

With highly interactive learning, the quality of each interaction is very important. The student with a human tutor is as before our guide to discussing such quality.

Weakly interactive modes such as pointing and multiple choice do not provide sufficient interaction. Much current computer material is unfortunately less interactive than older material, because of the recent heavy reliance on pointing as an interaction mode.

Interactive software can put the student in an environment similar to apprentice learning. The learner is not 'told' things, but learns by carrying out the suggested tasks with the 'master' (here the computer) looking over the shoulder of the student and only occasionally offering advice or asking appropriate questions as needed.

The key to quality student interaction with both people and computers is the use of the language of the student, discussed in the next section. Previous approaches understand this need in 'talking' to the student, but often not in the reverse situation, with the student talking to the computer.

5.2.4 Language and interaction

An important aspect of human interaction for learning, as in our student-tutor model, is the use of our most robust human communication and thinking tool, our native languages. These rich systems have evolved over thousands of years. They probably have played a major role in human evolution. They have affected our history.

As indicated, pointing and multiple choice are not suitable for highly interactive learning. They are of very limited use in learning, as compared to our powerful native languages.

Students in a highly interactive computer-based learning environment reply to questions from the computer in their native languages, without restriction as to the form of the reply. It is these questions, carefully chosen by the designers of the units, and the computer analysis of their free-form replies, that enable us to adapt to the needs of each student and offer individualized assistance.

We will discuss later our system for producing such highly interactive software. However, it is imperative to state immediately that we do not so far use the tools of Artificial Intelligence such as full natural language recognition; these may be useful in the future. Our development of highly interactive material uses technological approaches already available.

In another highly interactive mode, the computer responds to student questions. This is usually practical within a context of a learning environment. Such a question may lead to questions from the computer, helping the student to find the answer.

However, we expect most of the questions to come from the computer. Socrates 2500 years ago was asking the questions, not the students! This is in great contrast to standard courses today where interaction often implies that the student is able to ask a few questions. Students may not know what questions to ask.

There are many languages in the world, so if learning material based on language is to work for all parts of the world, we must consider the problems of moving material from one language into many other languages. Cultural factors may also be important at this time. Both output from the computer and input from the student are critical. A related problem is maintaining learning material already available in many languages; a change made in the program in one language must be propagated to the other languages.

5.2.5 Voice for input and output.

In highly interactive material, students use their native languages to communicate with the computer. In the past, including materials we developed at Irvine, keyboard input was the only method possible. Keyboards are not the world's most pleasant devices. We now have a better way of communicating, through speech. Computers can now recognize human speech.

A natural way for the student to communicate with people and with the computer is through talking. This enhances interactivity, as compared to typing and pointing. Science fiction has long recognized that talking to computers is the desirable way for humans to communicate with them, as we saw in the examples in the previous chapter. But so far, in spite of adequate inexpensive speech recognition software, there is little learning software using voice input available, and almost no good empirical evidence about the effectiveness of voice input in learning. Our attempts to get funding to carry on experiments in this direction have been unsuccessful.

Unlike the use of voice for dictation to computers, where we need a large vocabulary, no training of the voice recognition system for the individual student is needed for tutorial computer material. The speech recognition software is user independent for tutorial material. At any input, the program specifies important words in the student input, so the software for voice input can easily check for the input. Speech recognition is now practical and inexpensive for all tutorial learning with computers, although we need more experience.

Commercial software is available from several vendors, including Dragon Systems, Lernout and Hauspie, IBM, and Philips. Lernout and Hauspie now owns Dragon. New products are available frequently.

Full natural language recognition, pursued by artificial intelligence is not needed for tutorial material, as noted. It may be useful when it is possible.

5.2.6 Constructing knowledge

One major advantage of active learning is that it allows all students to create their own knowledge. Students can discover the laws of genetics in a way similar to how a scientist might, rather than reading or hearing about them. This happens in the Families Module in our Scientific Reasoning Series, described further later in this book. All students can make the discovery.

Experts in learning agree that such constructed knowledge is better than memorized knowledge told to the student. However, we have few materials today that allow and encourage individual discovery.

5.3 Cooperative learning groups.

So far, we have discussed interaction between computer and student. However, another type of interaction is also important in the learning approach discussed in this book, interaction between humans.

Students learn well when other students are working with them. Peer learning in small groups will be encouraged in future distance learning activities. Both local and electronic learning circles, arranged by the computer, are possible.

Research shows that groups of two to four are best in increasing learning and human interaction. Students engage in intense peer learning, assisting each other, stimulated by the computer questions. These may be local groupings or electronic groups may be established through computer networks, as noted. The computer can aid such cooperative learning, determining what students are at the same learning location, and bring these students together. It can even address individual students in such a cooperative learning group, as in a program about problem solving developed at Irvine. This dialog addresses students by their names, asking them to respond individually to questions.

We expect these cooperative learning groups to change frequently, depending on the progress of the learner. The computer program is building an informal model for each student and so can assist in the regrouping of students. Later we may know enough about learning to establish formal models.

The stored records, discussed in the next section, will be critical in making such decisions. Students who have already learned something may be able to help students just learning it. Adults might also be involved in this way, perhaps within families if students have functional families.

Thus, students have over many years the experience of working closely with many people, giving them practice in working cooperatively with others. This experience will be valuable in work experiences later in life, where working in groups is often important.

5.4 Storing student information

A human tutor has a large and increasing background of information about each student. But most computer-based learning programs today, and other types of software, store and use very little such information about the user. The computer is excellent for storing this information, with learning material designed with this in mind.

We discuss this further when we consider the process of development in Chapter 11.

In tutorial learning, we expect this saving of information by the computer to be a frequent continuing process with each student. This saved material can be inspected later by the program and used to make decisions about what is to be presented next to the student. This contributes to the interactivity and individualization of the learning material.

5.4.1 What is stored?

The designers of each learning module, working in small groups, make the choices as to what information to store. Some items are obvious. In each learning sequence, frequent information should be stored about where the student is, perhaps replacing previously stored information of this kind, so we will know where to start students when they return to the program.

An important type of information to save is about the learning difficulties the student is having, in the broadest sense. Designers often suggest such information. They will also see other information that may be useful later, and they store this. Formative evaluation may suggest other items to store and to use. As we learn more about learning styles, information of this type will also be very valuable.

Stored data is very valuable for research on student learning. This is a very likely important possibility for the future, with individual student privacy maintained. This may suggest other useful information to save.

This data from students will be an extremely valuable research base, as the number of students using the modules grows very large. We can gather the information needed to understand learning far better than we have so far. Many researchers might use this material if it is available to them.

Students are often entertained and surprised when the computer remembers something about past performance. They do not expect such long-range memory with the computer. This may change, as students become more familiar with this approach.

The designers need not be concerned with storing too much information. Something not used later can be removed.

5.4.2 Where is this information stored?

When we first begin to implement tutorial learning material the storage of student information is likely to be in a server for a local network. Thus a school, district, or state might have a server for its students.

As use becomes wider, storage will need to be available nationally or internationally. In many countries families move frequently, and the stored information should continue to be available. We may need some of this information lifelong.

A different possibility is the use of smartcards that would store the student information. These cards, for each student, would store the needed information. They have the problem, however, that students might loose them.

5.5 What material is presented next

Conversational interaction between people has a certain natural flow. Computer based interaction should also have such a flow in learning material.

Making choices in the program is a frequent activity. These choices can depend on both the recent inputs of the student, and on the stored records. Again, the design groups make all these decisions.

Formative evaluation will suggest additional useful information. We can modify the programs appropriately.

Several factors can determine what learning segment is presented next.

5.5.1 Computer-gathered data

A critical ingredient for making decisions is the student response to questions. In highly interactive material, these responses are frequent, offering a rich source of data for decision-making. Several recent questions might be involved.

5.5.2 Stored information

Decisions can also use stored information about what the student has done before. Again, this is a decision of the designers. We may eventually be able to build models of student learning that will help. One such possibility comes from an idea first developed about seventy years age. We will discuss this next.

5.5.3 The zone of proximal development

Lev Semyonovitch Vygotsky was a Soviet psychologist born in 1896. He died of tuberculosis in 1934. His ideas have only gradually been known to the world, in translations and edited forms. One idea that has received particular attention is the "zone of proximal development," described by Vygotsky and many commentators.

The quote from Vygotsky that follows was published in English in 1978, in *Mind in Society*. The Russian original in which the chapter appeared originally is in a posthumous work called *Mental Development of Children and the Process of Learning,* in 1935. The section we are concerned with is 'Zone of Proximal Development: A New Approach.'

Vygotsky says that 'the zone of proximal development is the distance between the actual developmental level as determined by independent problem solving and the level of potential development as determined by problem solving under adult guidance or in collaboration with more capable peers.' He continues:

> The zone of proximal development defines those functions that have not yet matured but are in process of maturation, functions that will mature tomorrow but are currently in an embryonic state.
>
> These functions could be termed the "buds" or "flowers" of development rather than the "fruits" of development.
>
> The actual developmental level characterizes mental development retrospectively, while the zone of proximal development characterizes development prospectively.
>
> The zone of proximal development furnishes psychologists and educators with a tool for development.By using this method we can take account of not only the cycles and maturation processes that have already been completed but also those processes that are currently in a state of formation, that are just beginning to mature and develop. Thus, the zone of proximal development permits us to delineate the child's immediate future and his dynamic development state, allowing not only for what already has been

achieved developmentally but also for what is in the course of maturing.

. . . .

what a child can do with assistance today she will be able to do by herself tomorrow.

Vygotsky concludes this section with negative comments about standardized testing, still relevant today. To our minds, Vygotsky was trying, among other considerations, to characterize the individual learning interaction. Note that he mentions problem solving, not rote learning.

This is a magnificent idea! There appear to be two important stages in this interaction with the student for using these ideas effectively for improving student learning.

5.5.3.1 Assess the current relevant knowledge of the student

The word 'knowledge' includes problem solving and other higher cognitive skills. Relevant knowledge, the "fruit," already matured, will vary from student to student and from moment to moment for each student.

In classes there are far too many students to find the status of each for each at moment. Each student is different, so the teacher typically aims at an average, and tries to add some individual attention. With highly interactive computer material, we can build a moment-by-moment view of what the student already knows, as discussed in this chapter.

5.5.3.2 Determine the zone of proximal development for the student

Knowing the current state of the student knowledge, we can then determine what that individual student is ready to learn, the 'buds' or 'flowers.' This is the zone of proximal development.

Knowing the zone, we can assist the student in learning. Eventually the zone may be determined empirically, but for now we must rely on the insight of skilled individuals, teachers, the designers of the learning materials. Again, this changes on a moment to moment basis in the student's life, further complicating the task.

It is not surprising to learn that this idea is not unique to Vygotsky. Other students of learning have come to similar ideas. David Ausubel begins his book, *Educational Psychology - A Cognitive View*, with a page that contains only this:

> If I had to reduce all of educational psychology to just
> one principle, I would say this: The most important
> single factor
> influencing learning is what the learner already knows.
> Ascertain this and teach him accordingly.
>
> Holt, Rinehart and Winston, New York, 1968
> Second edition 1978.

Both of these two stages depend on skilled people working with individual students. As the numbers of students in the world have increased it has become less and less likely that we can provide the numbers of skilled people needed, almost impossible in classrooms. However, we can with highly interactive computer-based tutorial learning, the subject of this book. The software plays the role of a tutor. More experimental information is essential.

> . . . parents, friends and peers . . . act as mediators
> who provided structure to the experiences of the
> child . . . In Vygotsky's terminology, effective
> mediators are sensitive to their child's "zone of
> proximal development"
>
> John Bransford, Robert Sherwood, Ted Hasselbring
> In Constructivism in the Computer Age
> Ed George Forman, Peter Pufall
> Erlbaum, Hillsdale, NJ, 1988

The first stage is to find where the student 'is' now, the current state of her or his knowledge, the fruits. We find this from the questions the computer asks the students and their responses. The design groups in generating tutorial computer-based learning units perform this task, if the design is successful as verified by formative evaluation.

Determining the second stage may be more difficult. The task of keeping within the zone of proximal development is a new one for most excellent teachers in the design group. They are not explicitly conscious of this task. We must rely on the experience of our teachers in working with students. Their knowledge may be only intuitive, not easily expressible in language. We need to understand better how good individual mentors, tutors, decide what the promising buds are. This is not an easy task.

Ideally, we should be able to advise the designers as to how to do this; perhaps we will be better able to give this advice in the future. Working in design groups may help, because teachers stimulate each other

How successful will tutorial computer-based learning units be in determining the zone of proximal development for each student at each moment? Will the tactics necessary be different in different subjects or cultures? These questions are empirical, not decided by argument alone. The approach is promising. We need to do the research.

5.5.4 Success in making decisions

Success in designing tutorial computer-based units, based on the factors just discussed, will depend on just how different students are. If every student is wildly unique, we cannot succeed, as noted. A friend, Arnold Arons at the University of Washington, suggested to me that the successes of highly interactive computer-based learning modules we have already seen indicate that while students are different, they are not vastly different in most cases.

This issue was raised at the beginning of this chapter. The fact that a few wonderful teachers succeed even within the classroom environment with 20 or more students also suggests that our chances are reasonable.

5.6 Mastery learning

In traditional education many students learn only partially, and some do not learn at all. Not everyone makes the highest grade, indicating that many people have not learned the subject as well as they could have.

A major advantage of highly interactive tutorial learning is that everyone can succeed. Benjamin Bloom and his students call this mastery learning. They showed, with extensive experiments in the Chicago public schools, that with human tutors everyone could learn everything to the mastery level. But we cannot afford such tutors for everyone. The computer gives us a way to accomplish this. We also consider this in later parts of this book.

With individualization and interaction, we can assure that all the students learn everything, to the 'A' level, mastery learning. If we determine, while the student is working with the learning material at a computer, that a student has not learned something, the computer can immediately try another direction, perhaps a learning approach more suitable to that student's learning style.

We can verify at frequent intervals in the computer units that mastery, success, occurs, for all students, not just at the large intervals that separate current testing. So with these new courses all students will succeed; they will learn almost everything, to the mastery level.

In such interactive material, learning and testing are intimately combined. They are fundamentally the same activity. This is a major difference from current educational practice. Much of the learning can take place during the assessment, and indeed this is highly desirable.

However, we can take a different point of view about this. We can say that the entire material is assessment. Every piece of instruction assesses the student. Previous and present assessments determine the flow of the learning material.

This approach is adaptable to modern interactive technology. It represents a full combination of learning and assessment, allowing us to achieve mastery. We discuss later how to organize the learning sequences for mastery.

Supporters of the mastery learning approach believe that everyone is able to learn a topic to mastery if given the learning material that fits their interests and learning style. Bloom suggests five outcomes likely to occur when students are required to learn each lesson to a criterion of mastery:

- They require less time to learn on later tasks,
- they are better motivated,
- they have a better self-concept,
- their achievement is predicted by the previous mastery task
- they experience less frustration while working on a task.

There are those that do not agree with Bloom and his supporters:

> ... the mastery approach may have a "Robin Hood effect": helping slower learning students at the expense of faster-learning students. Remember it is the classroom teacher who provides supplementary help in mastery learning. While the teacher is helping slower learners, faster learners may have to do 'busy' work, rather than go on to new units of curriculum. Just as

the legendary Robin Hood stole from the rich to help the poor, so mastery learning may take educational resources (teacher's time and attention) from faster learners to benefit slower ones.

This discussion is in the context of the classroom, where it may well be true. However, by introducing interactive learning material, built on the mastery learning approach and using the computer as the medium, the Robin Hood effect will be eliminated.

Assessment and learning will be intimately combined, with the purpose of assessment primarily to determine what learning material to present next to the student. Students will take different amounts of time in learning, whatever is necessary to ensure mastery. Some will move much faster than current students, as already suggested. This is particularly important in adult education, because of the very variety of students with many different backgrounds who would be involved.

Mastery will encourage lifelong learning, because successful learning, like all success in life, is enjoyable. It does not depend on threats, and learning success happens for all. All students succeed in a mastery environment.

Students can construct their own knowledge, in the tutorial learning paradigm, as contrasted to being told things. Students learn better by creating their own knowledge, rather than by being told things.

5.7 Motivating learners

Our final consideration in this chapter is that students should want to learn. This is a responsibility of the learning material, important for individualization.

5.7.1 Student motivation

Keeping students interested in the learning process is essential, so all the materials developed will consider motivation. Not all learning is easy or entertaining, so motivational issues are important; designers consider them, and formative evaluation establishes the motivational effectiveness of the material.

An important consequence of multimedia interactive learning is that we can keep students involved in difficult learning tasks. Our studies in public libraries, where students are free to leave at anytime, show that interactive material is self-motivating, if the interaction is of high quality. Interaction needs to be frequent; the student should be replying to meaningful questions at approximately twenty-second intervals, to maintain student interest.

5.7.2 Self-esteem

We want to increase students' self-esteem. This is possible in a mastery environment, where all students will succeed. Positive ways of addressing the student are important in this regard.

We want students to like to learn, encouraging lifelong learning. These effective components will be given careful consideration in the development.

Chapter 6

MULTIMEDIA IN LEARNING

A wide variety of media can be used in learning, including distance learning, such as print, lectures, conference sections, tutors, pictures, video, sound, and computers. Any one instance of distance learning will make choices among these media, perhaps using several.

Each can occur in a variety of forms. For example, with video we have broadcast video, tape, CD, DVD, teleconferencing, and the Internet, wired and wireless. With computers, we might have email, chat rooms, list servers, and interactive learning sequences at various levels of interaction.

6.1 TYPES OF MULTIMEDIA

One aspect of the distinct learning styles of different students is that various media will often be useful in learning. This will differ in effectiveness from student to student. We review the possibilities in this chapter.

As with all facets of tutorial learning, pedagogical factors should direct the decisions as to media and places in the learning sequences. In Chapter 11, we will see that this is one of the roles of the designers in our development procedures.

6.1.1 Text

A classical medium for learning is text, the written or printed word. It occurs in textbooks, in most web-based learning, and in student notes.

Writing is a relatively new form for communication, arising about 2500 or 3000 years ago. It probably occurred first in pictorial and character form, as in Egyptian and Chinese. The first languages that used a phonetic alphabet were perhaps Hebrew and Greek.

A problem for using text in learning is that unfortunately not everyone can read. This is true not only in the poorer parts of the world, with limited education, but also in developed countries such as the United States.

A conceivably even worse problem is that many people today, perhaps because of television, do not like to read. Many homes, even in wealthy environments, have no books.

Many readers also read in a sloppy fashion, not conducive to learning. When a large Sunday paper arrives, it is too long to read carefully. Therefore, for many reasons reading may not be a good way to learn for many students, although it works well for some. In the fictional accounts of learning we have seen, reading is critical.

6.1.2 Lectures

The lecture appears to have originated in Greece. Socrates made negative comments about this way of learning. It is a dominant mode in both schools and universities. Video of lectures is very similar in its effects.

Even with good lecturers, this learning media is boring for many students. If we visit almost any large lecture group, we find people sleeping. Even in small groups, this happens. This shows that for many students lectures are very boring.

Another interesting negative source of information about the value of lectures comes from examining student notes right after a lecture. They are frequently poor, reflecting little of what went on in the lecture. This examination is often a sobering experience for the lecturer. This example is a frequent one, even for 'good' lecturers.

6.1.3 Pictures

The use of visual information by humans far predates writing, as seen in such early efforts as cave panting. For many students this is an important form of learning. However, many teachers do not handle visual information well. They may not think visually.

Pictorial information should have some pedagogical purpose, not just be cute. Clip art on computers is seldom appropriate.

6.1.3.1 Computer Graphics

In the early days of computer use in learning, the only visual aspects of the programs were graphics supplied by the program. At first, these pictures were relatively crude, as judged by today's standards.

Most of the languages available at that time had no graphic capabilities. Further, early devices that would display graphics, such as the Tektronix 4010 storage tube series, cost more than non-graphic terminals, and the screens could not be selectively erased. Color was not generally available except on expensive terminals Some argued that graphic were not important for learning. One of us spent much time at meetings and consultation making the case for graphics in learning.

The graphics situation changed rapidly with the advent of personal computers. Even the earliest personal computers had graphics available. Now no one would argue that graphics are not useful for learning. It is odd, but technology often clouds our reasoning.

Early graphics on personal computers was often crude. Some examples from a dozen years ago are in our discussion of the Scientific Reasoning Series in the next chapter. Computer graphics have improved greatly since these early days, with better color and resolution. Their use in films and television commercials has been an important factor in this enhancement. Both still pictures and some animation, sometimes crude, have become common with computers.

6.1.3.2 Photographs

Learning programs can effectively use still photographs. Until recently, artists scanned ordinary photographs to put them in digital form. Now the availability of the digital camera may make the process easier. Pictures from digital cameras do not need conversion from analog to digital.

We can give an example from the work at Irvine where a photograph might be superior to computer graphics. In Heat, further described in the next chapter, the program places a pan of hot water on a table at one point. Text describes this and there is an illustration.

However, a few students persist in thinking that what they see is a stove, so the learning sequence does not work at that point for some students. We tried several times to draw a new table, but the problem persisted. Perhaps we could do better with the superior graphics now available, but a photograph of a table, or a very small video sequence of moving the pot from the stove to the table, would leave no doubt for any student.

Photographs offer a degree of reality generally superior to computer drawings. This can be important in some situations. An interesting example of the importance of giving an impression or reality is in the film designed by Charles and Ray Eames for a United States exhibition in Moscow. They wanted to show various United States structures - freeway exchanges, bridges, buildings, etc. They felt that a single scene of each of these types could be construed as a unique example, not characteristic of the country as a whole. Therefore, they showed dozens of photographs of each type, in a short period. The effect has been widely imitated, showing its success.

6.1.3.3 Video

Video is widely used and recommended for distance learning, but often in poor ways.

We have already seen several examples of the extensive use of video in distance learning. The Chinese TV University, the Stanford engineering courses, and the Annenburg CPB courses, mentioned in Chapter 4, were heavily dependent on video, used in a way similar to a lecture – in some cases they were primarily filmed lectures. These materials were not highly interactive, in the sense carefully explained in this book.

The video sequences went on for long periods, far longer than the maximum of twenty seconds between meaningful interactions suggested for highly interactive units. They are essentially lectures, talking heads and talking blackboards, with some additional approaches in the better material. Programs such as Nova use many different talking heads, with interesting visual material also shown.

Proponents of such non-interactive use of video are frequently promoting higher bandwidth Internet connections to support steaming video. We do not recommend this as a high priority for learning. It only makes long video sequences with no interaction more likely; if it can be done, it will be done. Further, such a direction further separates the already unfortunate learning gap between the rich and the poor, because greater bandwidth costs more.

Our Understanding Spoken Japanese project described in Chapter 7 shows a much more interactive use of video. We show only short video sequences of a few seconds at any one time, each followed with student interaction.

6.1.4 Sound

Another important media is sound. Headsets with microphones may help students in some environments to individualize interaction and the type of responses they receive. One student may prefer to hear a reassuring voice whenever she needs a little help understanding a concept, while another student may wish to 'strike up the band' whenever she has demonstrated understanding.

While films use music widely, its role in learning is not established. Perhaps it will have, in some places, motivational value. For example, it can stress a particularly important point.

Language use, vital in learning, is more complex. Many languages must be involved for worldwide use. It is estimated that in 2050 there will be 1.3 billion native speakers of Chinese, 556 million of Hindu and Urdu, 500 million of English, 486 million of Spanish, and 462 million of Arabic. Multilingual speakers complicate this situation.

For further information on languages, see
http://www.theatlantic.com/cgi-bin/o/issues/2000/11/wallraff.htm.

Many do not understand any of these languages, as these numbers do not add up to the nine billion we expect on earth in 2050. So to cover the world with a particular learning sequence we need many more languages, some spoken by only a small group.

The most common use of sound in distance learning is with voice. We need to consider both voice from the computer and voice from the student.

6.1.4.1 Voice from the computer

When the computer uses language, there are three possibilities. Perhaps all the language material is on the screen, already discussed under text. Alternatively, perhaps voice alone is used, with only pictorial material on the screen. Finally, both voice and text are together. These three possibilities are not mutually exclusive; several might appear in the same learning sequence. As with all media, these decisions are pedagogical, best made during the design sessions.

A probable situation where voice without text, but with pictures, will be necessary is with learning material for very young children. For example, an electronic headstart program, for children of two or three, must depend on speech for both input and output. Most of the students at that age cannot read.

In many situations, it will be useful to have both voice and text together. Many people, as we have noted, are poor readers, so having the material spoken as it appears on the screen, perhaps on a word by word basis, will help with the subject matter and probably with improving reading. This could be an option for the individual student.

An alternative, already seen in some educational toys, is that students can point to words they do not understand, and the computer can speak them.

The question of what voice or voices to use is important. In Chapter 4, we saw a science fiction discussion of this. It seems likely as our engineer suggested in *The Diamond Age* that computer-generated sound from text although feasible today is not currently of sufficient high quality for learning, but this might change soon. With human voices, it is useful, if we are to be gender independent, that we should have a mixture of male and female voices, and possibly even children's voices. We did this in the Japanese project mentioned.

6.1.4.2 Importance of voice input--talking to the computer

We mention several times that voice input is likely to be an important possibility for the future. Talking is a natural way to communicate for humans, much better than typing or pointing. In some situations, such as with very young or illiterate students, it will be essential.

With young children, there is some debate as to how effective voice input is, so we need more research. As we have emphasized, computer-based interactive learning requires no training for individual voices. Current commercially available voice engines are adequate for this purpose.

6.1.5 Computers

Computers are a new medium for learning. Some of its current usage fits into learning based on the information transform paradigm. Email is effective in distance learning, with small numbers of students. Chat rooms are also limited in numbers if they are to be effective. Listservs are also used. Computer-based teleconferencing is another possibility, but expensive and limited in numbers of students.

The primary use of the computer advocated in this book is as a vita part of highly interactive tutorial learning. This is a unique use, not possible in the other media used alone. Unlike the other media, the computer is inherently interactive, or can be if used properly.

The key is the level of interaction, already discussed in detail, involving both frequency of student-computer interactions and the quality of each interaction. This possibility guides us in preparing this book. We consider it in full detail in the following chapters.

6.2 Why media improve learning

We now address the question of why we use media in learning sequences.

6.2.1 Students are different

Student differences are already a common theme. One of ways that students differ is how they learn from media, and which media are best for which students. Eventually we may be able to predict this for the individual student, but now we can only have several media available, and then see which works best for each student. Having many media will mean that more students learn.

6.2.2 Motivation and Attitudes toward Learning

We hope, as suggested in Chapter 2 in our discussion of visions, that students should enjoy learning. This will increase desire for learning, both immediately and in the long range. Learning today should be lifelong, so we need people who want to learn. We return frequently to this theme in this book. It is particularly vital in distance learning. Media can be important in these motivational considerations.

In distance learning, without the usual threats of the classroom, student motivation becomes particularly important. Motivationally strong material encourages distance learners to stay at the learning task, even if it is a difficult task.

Motivational should be intrinsic to the learning material, not dependent on gimmicks such as loud music and violence. Interaction itself is a motivational factor; students stay at learning if it is responsive to their needs. Success, as in mastery learning, is an important factor in increasing desire for future learning.

We have long known from films and photography that media have motivational aspects. We should be able to use this in learning. The insights of experienced designers of media for motivation, such as those in advertising, should be valuable here.

6.2.3 Empirical determination of motivation

Motivation in computer based distance learning can be measured. We can put the computers in environments where there is no pressure to stay with the materials (no teachers, classrooms, periods, or exams,) such as public libraries and shopping centers.

Then we can store information about where students leave. The peaks in this information, where may people leave the learning program, indicate points where motivation is weak. We can redesign these sections, and test again. An essential factor in such weak areas we find is a low level of interaction. This process if described also in Chapter 11, when we consider the design process for interactive materials.

Chapter 7

TUTORIAL LEARNING

We have mentioned many times the concept of computer-based tutorial learning. It can lead, after the extensive development of appropriate learning units, to a distance learning system that can meet our goals in Chapter 2.

The tutorial paradigm for learning proposed here uses highly interactive multimedia computer-based learning. In this chapter, we develop this guiding idea in more detail.

> In the future, however, education will be organized largely around the computer. Computers will permit a degree of individualization - personalized coaching or tutoring - which in the past was available only to the rich. All students may receive a curriculum tailored to their needs, learning style, pace and profile of mastery, and records of success with earlier materials and lessons. Indeed, computer technology permits us to realize, for the first time, progressive educational ideas of 'personalization' and 'active, hands-on learning' for students all over the world.

Howard Gardner   Technology Remakes the Schools
The Futurist   March-April 2000   Page 31
The Disciplined Mind   Simon and Schuster NY 1999 Page 43

A primary problem in learning we have mentioned many times is that many students do not learn, either in conventional learning environments or in most distance learning environments. However, this does not need to be the case. Everyone can learn.

We have noted examples all through history of marvelous learning in which everyone learns. Some involved excellent teachers. Many of these teachers were tutors.

7.1 Tutors

Tutorial learning refers to learning with a highly skilled tutor and one student or a small group of students. The focus is on individual students, on learning rather than teaching. Learning is fully active for the student.

The primary consideration is the student as learner, rather than on authority figures giving information as in the information transfer paradigm. With very good tutors, people, this approach can provide excellent learning.

With a skilled tutor, learning is a highly individualized experience, with the tutor paying close and continuing attention to the learning problems of each individual student. A primary tutor activity in this interactive learning environment is to search for student problems, and to offer assistance for the problems discovered. Students will continue to work with the tutor until the material is fully learned. Peer interaction may also be important in the small group working with the tutor.

One critical aspect of this tutorial approach is almost obvious with human tutors. The tutor and the student or students talk in their native language. Our languages as mentioned are our most mighty intellectual tools, essential for the best learning. Another important aspect is that the tutor knows something about the student, learning status, problems, and desires, from previous sessions with the student and uses this knowledge in each session.

Although we have many examples of such good tutorial learning, it has never been the mainstream of learning. As the numbers of students has continued to grow, the idea of having everyone learn with good tutors has become impossible.

Tutorial learning with human tutors costs too much, and there are not enough good tutors. Only a fortunate few, typically from affluent backgrounds, could learn from tutors. Meanwhile, the methods we have used for learning for most students have become less effective, again partially because of number of students who need to learn.

7.2 Computer-based Tutors

Now the situation has changed, as we have commented. We have the technology and the knowledge to make tutorial learning available for all six billion people on this world

The computer, with programs developed by groups of excellent teachers, as described in Chapter 11, can provide tutorial learning for everyone, programs that work with the student until mastery. These programs can be used anywhere at anytime. They can be highly motivational, keeping the remote student at difficult tasks. However, very little such tutorial learning material exists yet; so more information is essential for progress in this direction.

Highly interactive tutorial learning with computers is a future vision for learning, not our current situation. We have described this approach in many papers, such as the EDUCAUSE Review paper for January/February 2000, available under papers at http://www.ics.uci.edu/~bork.

A computer-based tutorial learning experience will not compete with the best human tutors. We are not able to produce such material, and perhaps will never be able. That is not necessary to make major improvements in learning. Our competition is not the very few excellent skilled tutors available, but the learning systems that include most people, such as current schools and universities. Tutorial learning based on computers can be our learning paradigm for future learning, leading to major improvements in learning.

The advantages of human tutors will mostly be available in the computer-based tutorial form. Learning can start where the student is

> We doubt that the Quality of Instruction can overcome the effect of lack of the prerequisite cognitive entry behaviors unless the instruction is directly related to remedying these deficiencies or unless the nature of the learning task is sufficiently altered to make it appropriate for students in terms of the entry behaviors they bring to the task.

Benjamin Bloom
Human Characteristics and School Learning
McGraw Hill, 1973, page 109

The learning experience in computer-based tutorial learning would be highly interactive, conversational, in nature. Interactions would be in the students native languages, in both directions. Voice input now practical with commercially available speech recognition engines will enhance this interaction. Material will adapt to the needs of each individual.

7.2.1 Attaining mastery

A critical aspect is that the learning software can actively observe the students' learning problems on a continual basis, and can offer the appropriate assistance until each student masters the subject. This is critical for best learning.

It is not possible to determine these problems for each student in a typical classroom, as there are too many students. It IS possible and practical for computer-based tutorial learning, on a moment-by-moment basis. Computer stored records will play a role in this process, as noted in the last chapter. The zone of proximal development is an idea that contributes to this.

The learning activities would be adapted to the needs of each student. Students could stay with the material until mastery. Units will be highly motivational, encouraging students to stay at the tasks involved.

7.2.2 Comparing the two forms of tutoring

We have noted that humans will always be better for the student than a computer tutor. However, the computer-based approach would have many advantages over the human tutor.

Several striking differences between human and computer tutors are particularly important in bringing learning that is more effective to everyone in the world.

- Computer-based tutorial learning units could be available for very large numbers of students.
- With modern delivery systems, the student could be anywhere, not just in the location of the tutor.
- Students begun at any time, not tied to a schedule a human tutor might have.
- As mentioned, human tutors, working only with very few students each are expensive, impractical for mass learning. However, the financial situation is very different with computer tutors.

- Even with very expensive development cost (to assure excellent well-tested material for different types of students, leading to mastery) inexpensive delivery systems and very large numbers of students can lead to very low costs per student hour of learning. We can expect much lower costs, as well as much better learning, than in our current educational systems, after the new system is in place. Chapter 12 discusses the costs of learning.

Modern delivery methods make all this possible.

7.2.3 Delivery of Tutorial Learning

We review here briefly the questions of how such effective learning can be delivered worldwide.

Computer based tutorial learning could be distributed by CD-ROM, DVD-ROM, the Internet, or satellite, whichever proves to be most cost effective and easiest to access for most students. We need not choose a single method. Several delivery procedures will probably be used, particularly at first, depending on region or need. With these methods the cost for a student hour will decrease with larger numbers of students

Eventually satellite delivery seems most likely, when very large numbers are involved. Many regions of the world are presently inaccessible by other methods. Two-way satellite connections of this type are now commercially available. Perhaps the first vendor is Starband.

We need to investigate further delivering highly interactive material on the Internet or by satellite, since it is not yet happening. We need highly interactive delivery. There appear to be no major problems in providing such a level of interaction. Two-way communications between computers and students is essential in highly interactive material.

One possibility is that parts of the program could be loaded into the local computer, and the interactions could take place there, in a strategy similar to the Java Applet approach. This would reduce the amount of two-way communication necessary, helping the server and keeping the time between interactions to a minimum.

7.3 Is computer-based tutorial learning possible?

We have discussed a new paradigm for distance learning, computer-based tutorial learning. It depends on the creation of highly interactive software. We will later (Chapter 11) discuss one method for creating such learning modules. Nevertheless, readers may be suspicious that such material may not be possible, since they have not experienced it.

Very little existing material has the degree of interaction and the concern with student learning problems suggested in this book. Some might feel that technological developments beyond those available now are required for such tutorial learning material. However, this is not the case. Tutorial learning with computers is practical today. But developing the necessary learning units is a significant task.

7.3.1 Examples from fiction

We presented several examples of tutorial learning from fictional sources in Chapter 3. The most detailed is the learning that Nell did from her 'book', a computer. It occurs in Nell's homes and at other locations, distance learning.

It satisfies all the factors discussed in this book for computer-based tutorial learning. The primer is always sensitive to Nell's needs, and the learning results are spectacular, as judged by her success. The discussion in Chapter 3 does not give all the details found in Neal Stephenson's book. Sally in George Leonard's book is another example.

This is still fiction. But we have been developing material of the type promoted in this discussion for over thirty years at the Educational Technology Center at the University of California, Irvine. The computer technology was much more primitive than that available today.

The next sections describe several examples.

7.3.2 The Scientific Reasoning Series

We next consider programs in the Scientific Reasoning Series. This is a group of ten programs, twenty student hours for a typical student, intended to help students of about twelve to eighteen to understand the processes of science, how the scientist works. We developed them about a dozen years ago.

These tutorial programs are highly interactive, in the sense described in this book. From a student viewpoint, they resemble a conversation with the computer, a dialog. So they are tutorial.

Much of this development involved computers more primitive than those available today, starting with timesharing and then moving to early personal computers. The two programs discussed in the next two sections were initially developed on a Terak, with 64K of memory and no hard disk. The IBM K-12 group in Atlanta supported the conversion to IBM PCs, including the addition of color, and marketed the Series for many years. With interaction and tutorial learning, they satisfy all the requirements discussed in this book.

They show their age, however, with the graphics, not up to the standards of contemporary graphics. The programs do not use voice input, although that could be added. They could also be adapted for web use. We discuss Heat and Families from the Series.

7.3.2.1 Heat

The learning program Heat takes the average student about one hour, one of the shorter programs in the Scientific Reasoning Series. Students may not complete it in a single session. The aim is that every student should invent the concept of heat, starting with known information about temperature. It considers the learning problem of understanding the difference between temperature and heat.

The computer asks the student questions, and the student replies in free-form English. No pointing or multiple choice is used. No long passages of text appear on the screen. The first question, at the very beginning of the Heat program, is 'How do you measure your own body temperature?'

The screen is in color, not available here. The windows are different colors, shown in this illustration with gray scale.

Here is a screen showing the first student input.

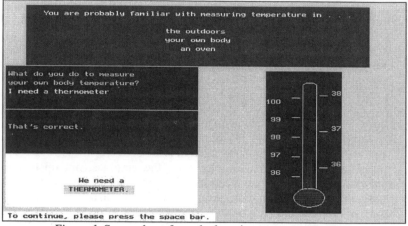

Figure 1. Screenshoot from the learning program "Heat"

Note that the first thing the student sees is a question from the computer. We do not start with a body of text for the student to read. This is exactly what the student sees. There is nothing else on the screen.

Students will almost always type (these programs as mentioned were developed before voice input was practical) something like 'use a thermometer.' If they do not, the computer makes the question a bit more specific and the student tries again.

Since this is the beginning of the program, the computer may need to convince students that they should type. So we watch, in the program, how long it takes the student to reply.

The picture here shows a possible response and the computer reply. The thermometer appears after the response. As commented, the graphics is not up to modern standards.

The next question asks if the thermometer will read the correct temperature if it is kept in the mouth only briefly, and the common reply is no. Then the next question is how to obtain a more accurate value, as shown in the next picture.

Here is the screen after the student has replied to this question. The box at the lower right appears only after the student reply.

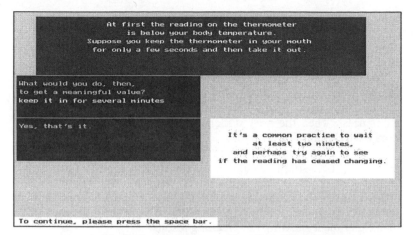

At first the reading on the thermometer
is below your body temperature.
Suppose you keep the thermometer in your mouth
for only a few seconds and then take it out.

What would you do, then,
to get a meaningful value?
keep it in for several minutes

Yes, that's it.

It's a common practice to wait
at least two minutes,
and perhaps try again to see
if the reading has ceased changing.

To continue, please press the space bar.

Figure 2. Another screenshot from "Heat"

This last question has several possible reasonable answers, all anticipated by the computer program. The screen shows one possible sequence. Again, we miss color. The reader will note that there is little on the screen at one time; this is as mentioned the entire screen, with nothing omittcd. In particular, we do not see any of the usual browser mechanisms, unrelated to the learning task, often found in today's online material. Chapter 11, in considerations of screen design, discusses the issues involved here.

The 'conversation' between the student and the computer proceeds in this form in the Heat program for about an hour. The object of the program as stated is to have the student invent, discover, the concept of heat, using kitchen physics examples, experiences the student probably will already have. We want the student to understand clearly that heat is different than temperature. All students are successful, some with unobtrusive help from the computer.

In some of the other highly interactive tutorial programs in the Scientific Reasoning Series, students discover important laws of science. We will next discuss a program concerning the laws of genetics. In another program the laws of simple circuit theory are the subject.

7.3.2.2 Families

This second example from the Scientific Reasoning Series is a discovery module, in which all students construct an important scientific theory, Mendelian genetics. The discovery procedure is like that of a scientist, although the activity, Families, does not attempt to recreate Mendel's experiments. This is a longer program than Heat, taking about two hours. Students probably will not finish it in a single session. The task is not a trivial one.

In the first sequence of eight, the student lands in a strange planet inhabited by critters called Nors. The first task is classification, finding how Nors differ from each other. There are only three differences. Then the next student activity is to find which pairs of Nors have children.

The remaining extensive task is the discovery of the laws of genetics of these simple creatures. As the student is performing online breeding experiments, the program is watching what happens and offers advice when necessary, just as if the student were working as an apprentice in the laboratory of a scientist. Here is the outcome of one such experiment showing the breeding results.

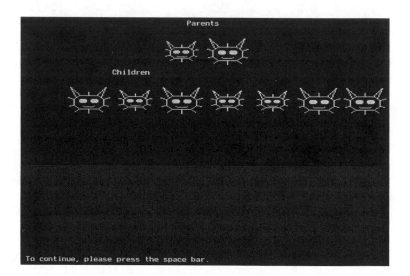

Figure 3. Screenshoot from "Mendelian genetics"

A complicating fact is that the evidence for the laws of genetics is statistical. That is, two pairs of seemingly similar Nors will have different children. So the task is not simple, but again all students succeed, perhaps with some limited help. The computer watches what the student is doing, and offers help when necessary.

Such words as genes, dominant, and recessive do not appear in Families. We strive for ideas, not vocabulary. However, students are prepared for this introduction of vocabulary. The ideas are there before the words.

7.3.3 Understanding spoken Japanese

A more recent example of computer-based tutorial learning developed at the University of California, Irvine, concerns learning Japanese in United States universities. The emphasis is on recognizing the spoken language. The modules use the interactive videodisc, making them difficult to demonstrate at present. If we were to prepare a new version, or continue the development, we would use DVDs.

Each module contains a video segment of about fifteen minutes. These segments are from Japan, so the language is standard contemporary Japanese. However, the student does not see entire module except possibly at the end of the module. To do so would violate our 'rule' concerning the frequency of interactions.

7.3.3.1 A typical activity

A typical sequence in these modules plays a small amount of video (perhaps five or ten seconds) and then asks the student what the conversation is about. As with all tutorial material, a free-form input from the student is accepted and analyzed. If the student knows the content, the learning program proceeds to another segment. If not, help at several levels is available.

Although we began with full language segments, we might eventually go down to presenting individual Japanese words and asking the student to repeat these. The aim is mastery. Note that with this strategy students do not 'learn' things they already know, so we are careful to minimize student time.

7.3.3.2 Language learning approaches

We developed several tactics particularly appropriate for language learning for these modules. Thus, the student finds where in a short video sequence people are discussing houses or another subject. Students have a facility like a videotape unit, allowing them to move back and fourth within the limited segment. The controls are on the screen, activated by pointing.

When students believe they have found the appropriate place in the video segment, they indicate that they have. If they are successful, the program may ask for another place in the segment, or may proceed to another topic. If not, the student receives appropriate help. Interaction helps the student learn.

This last approach seems very interesting for many other second language learning situations, such as English as a second language. In special areas we may well discover other tactics particularly appropriate to the area involved.

7.4 Summary of computer-based tutorial learning

In this final section of this chapter, we summarize the features of computer-based tutorial learning modules.

- The programs try to stay within the zone of proximal development for each student, as described in Chapter 5.
- The programs make a conscious and continuing effort to determine what learning problems students are having, by asking questions to the students.
- The programs offer help with these problems, perhaps at several levels and with different learning sequences.
- The programs check in each case to see if this help is effective, continuing until each student succeeds
- Evaluation and learning are intimately combined.
- Programs move at the student's natural rate, differing with each student.
- Programs frequently save and use student information.
- Individual student interactions are frequent, every few seconds.
- Interactions with the student are of high quality, frequently in the students' native languages in both directions.
- Evaluation involving large numbers of students of many backgrounds follows development. Improvement may follow.
- Evaluation includes motivational aspects: Does the program maintain the interest of the student?
- Programs are available eventually in many natural languages.

We need to educate more people,
educate then to far higher standards,

And do it as effectively and as efficiently as possible.

An American Imperative
Report of the Wingspread Group

Learning is . . . a necessity of life,   It's a consummation,
like eating or making love   . . . people are still half-starved
for learning.

James Cooke Brown
The Troika Incident, Doubleday 1970

Chapter 8

DELIVERY OF LEARNING

This book focuses on distance learning. It is our belief that the best form of distance learning is tutorial learning

Students can be anywhere and can learn at anytime, given appropriate delivery methods. An important component for such new learning methods is lifelong learning. In our rapidly changing world, we must continue to learn all our lives, as Peter Drucker points out. The learning enterprises must be managed, and perhaps credit or certification given. These issues are further discussed in Chapter 14.

As schools and universities decline as centers for learning, as they may, informal learning will become increasingly important. Tutorial distance learning can be delivered everywhere. All lifelong learning will be informal learning

The technology is evolving rapidly so the options we have today for delivery will probably be different tomorrow. This may be very helpful for the developers of highly interactive learning material. The limitation developers where faced with few years ago have almost disappeared but others have risen, therefore is important that we stay alert and follow the changes in technology and adapt to them as quickly as possible.

Our concern in this chapter is how the tutorial computer-based learning material will reach the student. Some things here have already been mentioned. In all likelihood several methods will prove to be useful. We need not pick one.

Our emphasis is on learning, not on a particular technology. The answer for delivery will depend on geographic location, and will most likely change with time.

8.1 Older delivery methods

In our discussion of distance learning in Chapters 3 and 4 we mentioned several older approaches to delivering distance learning material. This included mail, radio, and television, either live or through tape or CD-ROM.

None of these would be possible for tutorial learning, because they do not, allow the highly interactive format we required for tutorial learning. So we need not give them further consideration, except possibly as a delivery method for auxiliary material. We might, for example, deliver background material via video.

8.2 CD-ROM and DVD-ROM

A common mode for distribution of computer-based learning software is with disks, initially floppy disks. IBM distributed the highly interactive Scientific Reasoning Series developed at Irvine in this form.

CD-ROM had a much higher storage capacity, so what had once required many floppies could be on a single disk. Almost all types of software, including much learning software, are now distributed with this approach, although updates are typically distributed through the Web. Software stores are full of programs available on CD-ROM.

DVD-ROM is newer and has a much higher storage capacity. It is similar in physical appearance to CDs, and computer drives today are typically built to handle both media. So far this additional capacity has been used primarily for delivering films to homes. It clearly has the capacity for highly interactive learning. Unlike CD-ROM, with its limited storage, it can store many video segments to use in the learning activity. In at least one development at the Open University it is being used for such a purpose.

8.3 The World Wide Web

The World Wide Web receives tremendous attention. It is often recommended for delivering everything. The delivery method that receives the most attention now for distance learning is the Web, using the Internet. Online learning usually means through the Web, not just online to a computer. Some talk as if this is the only possibility for the future.

We will just call it the Web. Primarily wires and cables are the transmission media involved for Web delivery. Costs determine speed of transmission of information, so there is a digital divide concerning Internet access, for this and other reasons, between the rich and the poor. Slow connections are through phone modems, while faster connections are typically through Digital Subscriber Line (DSL) or cable networks.

Local distribution can be wireless, as in the NetSchools environment. This wireless distribution has many advantages, allowing a computer to be used anywhere in a school. We will discuss satellite access to the Web in a separate section, as many of the factors are different.

8.3.1 Web criticism

While the Web is often praised as an educational media, it also has its detractors.

According to Peter Goodyear, Lancaster University, United Kingdom, and he wants to be quoted,

> The World Wide Web is the biggest disaster to hit technology-based learning since the introduction of the personal computer.

To some extent this is true, as the introduction of the World Wide Web and its capabilities has caused a pause in the development becuase people have been busy learning how to take advantage of the Internet. . So far not much has been done; as we stress there is very little interactive tutorial material developed and available..We are now in the year 2000 and we are almost in the same place in development of new learning units as we were before the Internet was available to the general public. We have little effective learning material available. In fig.4 Peter Goodyear draws up the evolution of time and the progress of media richness, interactivity in learning material and pedagogy. We can see that in 1995 we are more or less in the same development phase as we were in 1985. He predicts that by the year 2005 we will be even worse off then we were in 1985.

But, what has changed is the growing number of people involved. The amount of money given to this area of research is far more than ever before. Therefore we disagree with Peter Goodyear's predictions and think we will see a booming effect of the use of interactive learning technology and a progress in use of pedagogy.

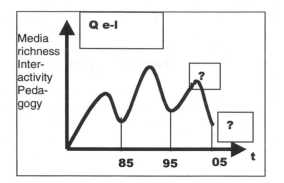

Figure 4. Goodyear's estimation of the effects of the WWW on learning technology

We will soon return to this issue of problems in distributing learning material on the Web.

8.3.2 Web advantages

The Web has some clear advantages, but not necessarily unique to it. It has a very wide reach, but only in the more affluent parts of the world. Many already have computers that can access the Internet, but again there is a digital divide here.

Since the material is stored in computers, it can be in principle changed rapidly for all users. This is often mentioned, but it is not clear how often this is, or will be, done. We need a systematic way to handle this situation, such as that at the United Kingdom Open University, but this is missing with most current learning software.

8.3.3 Current problems with Web learning

These problems are not fundamentally problems with the World Wide Web, but with the online learning materials on the Web that have been developed so far.

8.3.3.1 Units only work for small numbers

A major problem we face in learning today, frequently emphasized in this book, is increasing numbers of students. This is a global problem as we now have 6 billion in the world, predicted to be over 9 billion in 2050. Predictions from the United Nations suggest a rapid rise in the medium age, also affecting our need for learning, and affecting the type of learning needed. Adult learning will be far more important than at present.

Many people have inadequate learning available, in every country, even in developed areas. We have a billion illiterate adults in the world, for example, many even in wealthy countries such as the United States. For many students little or no formal education is available. Students from poor families, worldwide including wealthy countries, have inferior education opportunities.

The approach at many universities now, seen in Chapter 4, is certainly not the only form of distance learning. Current Internet learning activities are not based on empirical data about learning; they appear to be based on expediency, and a false sense of economy.

Many different possibilities for distance learning other than current online courses are possible. This myriad collection of distance learning types should be considered before deciding on a single, or several, systems.

Experimental studies with large numbers of students are required to make rational choices as to the best form of distance learning for a given situation. These studies do not yet exist. Mostly, they have not even begun. We will discuss such studies later.

8.3.3.2 Lack of adequate interactions with students

Students need individualized help for effective learning. Current learning, both in classes and on the Web, often assumes, as we have seen, that the central task of education is the transfer of information to the student. But this view is not adequate for the new century or for many students and for many areas. It does not easily lead to development of higher cognitive skills, such as problem solving and creativity.

We need in learning to be concerned with what the individual student does not know, and what problems the student is having. We can find this with frequent high quality interaction with each student, but current Web materials do not have this level of interaction.

As discussed, frequent means every few seconds, and high quality demands that the interaction should be, in **both** directions, in the student's native language, our must powerful tool for communication. This means very little pointing and multiple choice on the part of the student. The student's language is very important in this interaction. The computer can ask questions, looking for student problems, and students can reply in free form.

To achieve this on the Internet, we need not be concerned with rapid two-way communication between student and the server computer. Perhaps the best strategy will be to download chunks of the code, perhaps entire program segments, to the local computer, and have the frequent interactions take place locally. This is similar to the Java Applet strategy already in use for other purposes.

We need also to consider interactions with other students, peer learning. This receives little attention in current online courses. Peer learning is an important component of learning. The learning units should stimulate it, bringing students together. Groups of about four are best, we believe, perhaps arranged electronically.

8.3.3.3 Learning is not available for many

Learning, particularly complete learning, is not always available for many students. This is obvious for the billions of people on earth who have never used a telephone, and for the very poor billions.

But it is even true in the developed countries. In spite of all attempts at equity in education, the poor are neglected, as are women. Current online learning does little to improve this situation.

8.3.3.4 Insufficient storage of student information

Skilled human tutors start a session with a student with considerable previous experience, and they make use of this information to guide the tutorial situation. Current Internet systems store limited student information, usually only to show overall progress and determining grades. But computers can store much more detailed information!

If we are to improve learning for all we want extensive records for each student, gathered on a moment by moment basis as learning takes place. Information about student learning successes and problems is particularly important. This stored information should be used, along with recent student responses and other information, to make decisions about what learning material to present next to each student. An important clue to what is needed for making this decision about what learning materials to present next comes from Lev Vygotsky's concept of the zone of proximal development, suggesting what the student is now ready to learn.

8.3.3.5 Many students do not learn with existing materials

We need learning systems in which ALL students succeed, learn to the mastery level. Learning is necessary for individual happiness and for societal progress. We cannot afford to waste talent in the new century. Evidence indicates that mastery is possible for all in tutorial environments.

But the current online learning materials do not help all students to learn. Many students drop such courses, and other show only partial learning. Many are bored. Since these courses imitate existing standard courses that have these same problems, this is not surprising.

8.3.3.6 Learning is too expensive

Present online material, with twenty student groups each with an instructor, is far too expensive for today. Further, it does not scale easily to much larger numbers. We need to consider new possibilities for learning that have more reasonable costs. These questions of cost cannot be ignored.

8.3.3.7 Insufficient consideration of lifelong learning

Most of the online material developed is based on current university courses, as we have noted. Current systems of learning focus primarily on students from about six to twenty-five years.

But the demographic data indicates that the center of learning is soon to move forward. Even today, the rapidly changing world continually demands new skills and new thinking, as we grow older. This trend will continue and accelerate. So we have the challenge of meeting this new need.

8.4 Communication

Communication is a vital part of learning. According To Terry Mays and others, learning is a social activity.

> Learning takes place in a social context. The nature of that context can have a significant impact on individual's approach to learning. The most important

social factors in education are usually the nature of the assessment and the attitudes towards learning of peers, teachers and employers (if the learning takes place in a work setting).

As a part of the learning environment the student will have access to services that support communication and build "virtual communities" around certain learning objectives. The Community Learning Service is an idea of service like that, a generic framework that aims to meet the needs of individual users who want to support and develop learning dialogues over a distance. It offers a range of different tools and communications methods and provides assistance in selecting and appropriate configuration depending on the nature of the community. The service encourages additional learning activity by providing directories of subscribers who are willing to enter into new learning relationships.

The service supports both individual users and communities. Individuals are provided with personal note books, diaries and search agents accessed via a personal learning portal, while communities of 2 or more users are supported with a shared environment that provides tools, storage space, joint e-mail, joint search facilities. Each community also has diary facilities, meeting and communication management tools.

When learning over a distance there is a tendency for students to feel isolated and not be a part of a group. This is why a communication with group of students mastering the same material could be important to raise the student's interest in the subject.

8.5 Communication media

It seems likely to us that the distribution method for highly interactive tutorial learning will eventually be primarily through high-bandwidth solutions, either with cable networks, DSL solutions, fiber optics or satellites. This may or may not be with the Internet, or with other Web standards. The access to these solutions depends on the location of the learner.

8.5.1 Highly populated areas

Today many people in the world, living in highly populated areas have access to high-bandwidth solutions, such as cable modems, DSL solutions, and even fiber optics to their homes. These services are expensive today, but the price will come down, in the coming years, because of the high competition on the market. This medium will be the highway for distributing our highly interactive tutorial learning to learners living in highly populated areas.

8.5.2 Rural areas

For people living in rural ares the options are not as many. It seems likely to us that the distribution of highly interactive learning material to the rural areas will eventually be primarily through satellites.

8.5.2.1 One way communication

Until now satellites have been little used for distribution of software, such as learning software. Most of this has been transmission only from the satellite, as in video delivery. If communication was required from the local computer to the network server, phone lines were used. This meant that the bother of arranging both types of connections was necessary, and that return communication was relatively slow. It also means that we must be in a part of the world where phones are available, and not heavily used for other things.

For tutorial software, we need to get information about student performance back to servers. So the one way satellite approach presents problems.

8.5.2.2 Two way communication

The use of satellites in both directions, as a practical thing, is a new development. No phone connection is needed for the return signal.

This service is now available from Starband, initially called Gilat-at-home, working with several other companies including Radio Shack, Echostar, and Microsoft. The antenna needed is very similar to that for television transmission from satellites. The costs, so far, are compatible with those for other methods of receiving Internet signals.

Other companies, including Hughes Direc-PC and Wildblue, are expected to compete for this market.

8.5.3 Advantages of satellite communication

As indicated, we should be able to cover the earth with a mix of high-bandwith solutions, bringing excellent learning to everyone. As emphasized this may or may not use the Internet. The problem with the Internet today is that the "quality of service" can not be guaranteed. As we would not want commercial competition for the broadcast time, and we can have our own standards, separation from the Internet may be desirable.

Chapter 9

LEARNING AND ASSESSMENT

Assessment is an important component of education. We have already suggested that assessment should hold a special role in tutorial computer-based learning. In this chapter, we explore assessment in more detail.

This chapter uses a paper by the authors and David Britton.

9.1 WHY DO WE ASSESS?

We first ask why we access students. The following reasons are relevant. The attitudes one finds toward assessment are dependent on underlying philosophies, often independent of educational content.

9.1.1 Assessment for grades

Assigning grades is the traditional role of assessment, placing the students in a class along some spectrum from good to bad. Such grades comparing students with each other are dominant in schools and universities. Graduation with good grades is too often the major goal of many students. Learning is not given such a priority.

A few universities have avoided assigning grades, but this is very rare. Initially the University of California, Santa Cruz, gave no grades; it replaced them by instructor-written reports for each student. This led to problems with graduate schools. Reed College has grades, but students do not learn them until graduation. The idea is that students should focus on learning, not grades. But the advisors know the grades, and so can 'advise' students.

However, the situation is different away from schools and universities. Most adult and informal learning outside of schools has no such grading system.

A problem with tests for grades is the inevitable student negotiation for more points. If a student misses a 'B' by one point, he or she may feel obligated to make up the deficit through talking with the instructor. If students are assessed incrementally as they learned and given immediate assistance with their problems, there would be no need for grades and no after-test negotiations.

9.1.2 Assessment for future education

Grades are often an admissions criterion for the next step in learning - getting into a university, getting into graduate school – along with exams such as the Graduate Record Exams (GRE) and recommendations.

9.1.3 Assessment for jobs

Students and parents often believe that grades will affect job possibilities. This may be true for the first job, but after that performance is probably more important. There is some evidence to show that grades do not correlate with success in life, as measured by a number of criteria.

9.1.4 Assessment for aid to students

Another reason for assessment is to provide responses to the student, helping with problems and showing need for further study. Some teachers consider this a major reason for testing.

This student advice is often poor with testing as it exists today. There is often a sizeable gap between the giving of a test and students getting information back. Therefore, students forget what decisions they made. Conventional testing requires students to wait at least a day or more likely several days to learn the outcome of an exam.

For many students, there is anxiety concerning how well they did until the examinations return. Some feel that for anyone that knows the material well may brush this aside. But most students wish to get their tests back in a timely fashion.

Often the graded test contains little information other than marks on individual items. Only in small classes is there likely to be much instructor advice. The situation is even worse for machine graded multiple choice quizzes, where the only information available to the student may be an overall grade.

A class usually moves in a lock-step fashion, paying no attention to whether the individual student has learned the material or not. After all a few students are learning, the 'A' students! The attitude in the typical classroom is that there is something wrong with those students who are not making 'A's, that they just are not trying hard enough or even worse lack the intelligence. The notion that they might be having real learning difficulties due to their backgrounds or other factors cannot be taken into account in a fixed-pace course. The notion that all students can learn is foreign to many teachers and professors, as they have always been in graded classes.

Another factor that interferes severely with giving assistance to the student based on assessment is the widespread use of multiple choice exams in the United States, already mentioned. We discuss this unfortunate tactic further later.

9.1.5 Assessment for parents

Assessment for giving information to parents and other adults associated with the student is very closely related to the first aspect of assessment we mentioned assignment of grades. Students and parents as well see the grades.

Closely tied to this is the question of threats to the student. Most schools in the United States no longer allow corporal punishment. The psychological threat of grades replaces the physical threat. The implication is that not only will assessment affect the grade, but also it will make your parents unhappy with you and will probably ruin the rest of your life! Teachers frequently make such threats, although the effect on the remainder of the student's life is dubious. Films often portray teachers unfavorably in this regard.

9.1.6 Assessment for judging teachers

A widespread use of assessment is in judging teachers, deciding if they will continue their jobs. Such 'high stakes' testing motivates educators to concentrate their teaching toward the material that is assessed. There are examples of teachers cheating to improve overall class grades.

9.1.7 Assessment for evaluating educational systems

Much recent literature about assessment has been concerned with assessment as a method of evaluating the school systems. International testing is now widely done. There is a general belief based on this testing that the United States system compares poorly with that of countries that compete economically with us, such as Japan.

The danger is that this process can distort the schools. According to Paris et al., the Stanford assessment, common in the United States, does not match well with the curriculum of a school the authors were developing, based on authentic assessment. Teachers found themselves 'teaching to the test' and not covering material that they felt matched the goals of their curriculum. Many teacher report informally similar results.

Although much of the recent emphasis has been on the schools, universities are under increasing scrutiny everywhere. Thus, the recent United States Wingspread report is very critical of the state of the higher educational system. However, assessment of universities is difficult, in any empirical sense.

9.1.8 Assessment for structuring learning

In this section and the next, we discuss the most important reasons for assessing students' progress. Assessment is part of learning. This is related to aiding the student, already discussed.

In a flexible learning system responsive to individual student needs, like the tutorial learning discussed in this book, frequent decisions are made about what learning material the student sees next. This decision ideally should depend on detailed information about student understanding of what has been happening, using recent student inputs, stored student records, and possibly information about the learning styles of the individual student. The zone of proximal development may also prove to be important.

This use of assessment to guide future student learning is rare, because in our existing learning situations all students are presented with the same learning approach and materials, at a fixed pace. However, such assessment becomes vital in a highly interactive tutorial environment combining learning and assessment.

9.1.9 Assessment for mastery

If a student is learning in a highly interactive tutorial computer-based environment, we can use assessment to determine what the student needs further help with, and then immediately proceed to offer such help. Therefore, with several types of learning aid available, we can bring all students to full mastery. That is, we can expect all students to succeed in learning; true democracy in education.

Grading is unnecessary and inconsistent in a system that plans on success for all students. Instead, students' progress toward mastery of the agreed upon learning outcomes is evaluated by comparing their work to the learning objectives for any given subject or school year, as determined by the designers of the learning units. Students' work and accomplishments are compared to a standard of performance, as opposed to other students (a criterion-referenced or mastery approach). Students do not compete. All students are successful.

9.2 TYPES OF ASSESSMENT

Various kinds of assessment are possible. We give only a brief discussion, not complete

9.2.1 Assessing for memory

A very poor form of testing is assessing only for memory. Of all the results gained from assessment, it is the least interesting It ignores such important intellectual tools as problem solving, intuition, and creativity. Nevertheless, it is a dominant type in both schools and universities today, because it matches the information transfer paradigm for education. It may also be very prominent because it is the easiest way of testing.

Knowing how to find information is more important than remembering information, increasingly in our society. Further, people forget material memorized for the test (cramming).

One alternative is to give open book examinations or allow a page of notes, or all notes. But this seldom happens. Examining most existing exams would show why this is not common, because of their reliance on memory.

9.2.2 Multiple choice

Multiple choice, already discussed briefly, is a very poor assessment form. It departs from the real world; we seldom encounter situations with only four or five specified choices. Most people would not like to have a brain surgeon whose skills have been show only by multiple choice. It allows students to pretend to knowledge they do not have.

One needs only to examine the extensive literature on 'how to take tests', almost all directed to multiple choice tests, to see how unsatisfactory this approach to testing is. This literature might be described as, 'How to do reasonably well on a multiple choice test without knowing much about the topic'. The approach is to advise students how to do better at guessing the answers. This literature is often successful in improving test scores. But the skills that students learn have little to do with learning in the subject areas.

Furthermore, student attitudes about multiple choice are negative, as seen by the common student description of this procedure as 'multiple guess.' This is not intended as a favorable description! Except for a few students who are skilled at multiple choice, and who do not want to extend themselves, most students do not like this form of testing.

Another problem is the difficulty of writing multiple choice items. Examination of professional groups specializing in multiple choice shows the difficult of writing good multiple choice questions, even with major funding and expertise to help; one of us worked for years on the Graduate Record Exam in Physics.

Multiple choice is the 'work of the devil,' unsuitable for humans!

9.2.3 Fixed time examinations

Working carefully and fast are prized skills, in the view of many teachers. But what about students for which these are mutually exclusive? Are they doomed to failure? Would it be unfair to give more time to the students who need it? Is fast performance that important?

In a conventional classroom situation, giving some students additional time might be unfair to others that need it as well, but cannot take advantage of it because of scheduling and time constraints. It would also be unfair to the teachers who must monitor the tests.

However, when given enough time, students that would do poorly on fixed time tests may be able to demonstrate understanding. One of the authors recalls a personal incident when there was not enough time to finish a calculus exam. At the end of the allotted time, the author conveyed this to the professor. The professor's reply was, "We don't want you to reinvent the wheel." The author knew the material, yet what if the "wheel" had been reinvented. Would not the purpose for 'king the course, to learn the material, have been satisfied?

Perhaps in some situations speed is important. But often in real life, it is not.

9.2.4 Authentic assessment

A very popular term today is "authentic assessment". But its use differs from person to person.

If you wish to gain authentic assessment (as well as gauge performance), you must pay attention to *the context of production* of the material. Many performance assessments - timed, with contrived problem scenarios - can hardly be called authentic. Additionally, authentic assessment must be concerned with the content assessed. It must address the needs of the curriculum while serving students' learning endeavors outside of school. This is not an easy task, but is worthwhile.

Authentic content can help students to connect ideas. While it is desirable for students to learn to work well in groups, there are no assessment measures for groups. Hence, individual assessment of knowledge and skills learned in groups may not be authentic.

A popular form of authentic assessment in the United States and elsewhere is based on student portfolios, collections of their work over some time. This idea can also be used in computer-based learning systems.

9.3 Combining learning and assesment

There are major advantages in combining learning and testing, realizable in a tutorial environment with either human tutors or computer-based tutors. We review them here.

With learning and assessment combined, there is no need for the word "test." Why test, in the conventional sense, at all? Why not build knowledge by testing incrementally and frequently, adjusting the learning dialogue to suit the student until mastery?

Students engaged in Socratic dialogue with one-to-one tutors, either people or computers, are continually assessed, The tutor tries to gauge comprehension and decide, based on the tutor's understanding of the student's knowledge and skills, what material should be covered next. Students will get exactly the help they need, as they need it. There is not the wasted time of the conventional classroom.

9.3.1 No tests!

The advantages of a learning system in which there are no apparent tests are great. Testing may still be happening, but the students do no see any tests in the conventional sense.

Most students do not like tests! More often than not, a class groans when an exam is announced. Tests are a source of anxiety and stress, even for the better students. Much student attention is concentrated on tests. A common student question is 'will that be on the test.' If the answer is no, the topic will be dismissed by many students. For some students, even bright ones, examinations are a disaster. Additionally, assessment may stifle creativity.

In courses at many universities, there may be only two tests in a quarter or semester, with little other basis for a grade. A bad day, such as a student illness, can mean a bad quarter grade. The test-taker's transcript will reflect a grade indefinitely.

Some may say that if the student knows the material, the test will still turn out all right, even on a bad day. The student will remember all of the test-taking strategies, such as not dwelling on any one question, attempting to get the easy points first, and making sure to get the big points. Is this what we want to teach our students? Are test taking skills, often specialized to multiple choice, as valuable as the material that we are testing? If not, then why do they affect the outcome of the test so much? If so, why? Because the lecture/ test cycle is so pervasive? Would test-taking skills continue to be valuable if there were no tests in the conventional sense? We believe the answers are obvious. We all know that students do have bad days.

9.3.2 Students always succeed

Why do we expect some students to be mediocre and still others to fail? Perhaps it may be argued that students having trouble can be 'put back.' However, most of these students will not need to repeat an entire year or course. Perhaps only a minor amount of work can bring the students up to date.

Students with missing or 'buggy' knowledge should be able to address only their weaknesses. What is the point of an educational system if not to ensure that students learn? What is the point of learning if not to mastery? Students working to the mastery level may need varying amounts of time to get there.

However, self-paced, individualized, one-to-one tutoring will give every student the chance to succeed. Students can avoid the stigma associated with 'flunking'.

Bloom and colleagues found the average tutored student was two standard deviations above the average of students with conventional instruction. Computer tutors can provide high quality interaction in the form of Socratic dialogue.

One of our goals for learning in Chapter 2 is success for every student. We can attain this will the learning system presented in this book.

9.3.3 Less-threatening learning environment

Computer-based learning systems that combine learning and assessment are not judgmental, ignore gender or other differences, never get tired, are infinitely patient, and do what the designers tell them. Additionally, they will not put someone 'on the spot,' embarrass, or pay more attention to some students over others. Fair treatment, not dependent on the whim of the teacher, is available for all students.

9.3.4 Assessment as it should be used

Assessment is not as a weapon to decide if students pass or to threaten teachers' jobs, but a tool to learn about gaps in learning. It should provide students with the assistance and encouragement they need for learning to mastery.

Tutorial computer-based learning systems may facilitate assessment at several levels. First, a student's knowledge and skills can be assessed, for example, through traces of a student model. Second, systems can facilitate metacognition, students' understanding of their own thinking. Finally, computer-based learning systems can provide 'meta-assessment'. We can find evidence of self-assessment in addition to assessing a student's knowledge and skills. Students can be helped to become more aware of their own knowledge.

With computer-based learning systems assessing as students learn, grades are irrelevant. Every student, learning to mastery, is automatically an 'A' student! All, students, from the slower learners to the brilliant, should and probably will achieve mastery, working at their own paces and receiving individualized help. Students should value learning, not grades.

Parents can still be kept informed and educated to the notion of mastery. Reports concerning students' progress are always available. They can be available online.

Teachers, in environments where they still exist, can see who needs individualized attention and concentrate on helping those students.

9.3.5 Increasing student motivation

Motivating educational material captures and holds attention, gives the learner confidence that success is likely to be achieved, and makes possible results that the learner is satisfied with. Students who succeed are surer of their abilities. Combining learning and assessment in tutorial learning lets us accomplish this. Success leads to further success.

Course material should provide high quality interaction in domains real to the student. Additionally, students can have a choice of methods and content of instruction. They need not be constrained to pre-specified paths through the material. High quality interaction with a friendly non-judgmental personal tutor giving immediate help is surely more motivating than passively listening to a teacher lecture. Computer tutors can adapt to the student-and unlike human teachers-can change personality at the student's request.

Students may receive progress reports or graphs indicating how far they have come. Reports and plans in the form of Pert or Gantt charts can help students and adults budget their time. Perhaps graphs of anonymous students could be prepared to allow students to see how their peers are progressing. Students could possibly locate a pointer on a continuum that indicates their personal progress with respect to other students, in each area of learning. We can avoid the introduction of undesired competition.

Chapter 10

STRUCTURES
FOR TECHNOLOGY BASED LEARNING

In this chapter, we discuss ways to organize learning, including possible structures for tutorial technology-based learning. What kinds of interactive structures, using modern interactive technology, are possible to assure mastery learning for all students?

The structures that we are interested in are those that combine, either partially or fully, the learning and assessment activities, as discussed in Chapter 9. But we begin with the traditional approach to organizing learning.

10.1 Traditional courses

A learning-testing strategy dominates traditional learning organizations, such as in schools and universities. Most of the time is in learning from lectures and books, in the information transfer paradigm. Testing occurs infrequently at periods of weeks, and it is only then that the teachers and students see what has been learned. Testing, typically based on memory, is primarily for assigning grades. Figure 1 shows this process. Almost all of these courses have instructor-controlled content.

Traditional courses in schools and universities may include the computer and its related technologies in the classroom experience. This is typically a minor addition, such as electronic mail, to a course based on information transfer.

Online web-based courses so far almost all follow this same approach, although there may be uncertainty as to how to give the tests. The problem, often discussed, is verifying who is taking the examination when it happens at a remote computer.

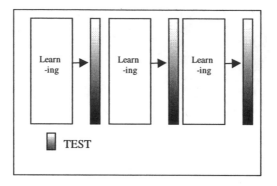

Figure 5. Segment of conventional course

Traditional courses are usually linear in structure. All students see largely the same material, except possibly for projects done in groups. Testing is based on memory, typically. In the next organization, different students may see different material.

10.2 Non-linear learning

One possible organization for learning is non-linear. Most courses today are linear. However, in a non-linear or hyperlearning segment, students can each follow different paths. From one set of learning material, many paths are possible. Different students may not see the same material.

Early examples of this type were hypercourses or hypersegments. English and biology courses at Brown University were elaborate experiments in that direction. Several elaborate hypersegment programs, making heavy use of media, were developed with support from IBM, Columbus, and Illuminated Books and Manuscripts. All these examples depended on pointing, so the quality of interaction was low. Due to the expense, probably, they produced few such units, and the material had limited use. There seems to be no current production of such sequences.

But high quality interactions, not just pointing, are possible in non-linear segments. The program, based on student performance may make these variations in the order of content presentation. The program may, for example, find that additional mathematical study is need at one point in a physics course, so will present the necessary module or modules to the student before returning to the physics sequences.

Another possibility is that the students have choices. We mentioned our strategy in a beginning physics course, where students had six possible paths through the material.

10.3 Mastery organization

We have frequently discussed in this book the goal of everyone learning, mastery. Success is the goal. In the following sections, we examine several ways of organizing courses for mastery. These might have either a linear or a non-linear arrangement, so this section complements the previous section.

Experience with mastery learning suggests that if a student does not learn something with a single approach, a different approach, perhaps quite different pedagogically as well as in media usage, may be necessary. So mastery based material may have many units that are concerned with the same topic. Data gathered shows which approaches are most effective with which kind of student.

10.3.1 Keller plan mastery course

Early organizations of mastery courses, such as seen in the Keller plan, about thirty years ago, break the material into small units typically requiring about one week of study for an average student. The student has information about the contents of the unit and available study material. There is a mastery goal or goads associated with a module. Both tests and learning material are required for the Keller Plan.

The computer was not involved in initial experiments with the Keller plan. Physics was a common area. Droves of teaching assistants graded tests in large courses. Students often felt that grading was unequal and unfair. This lead to the computer approaches discussed here.

10.3.1.1   Testing in the Keller plan

To see whether someone has learned the material or not there is a test, a series of questions presented to the student online at the computer. As is already indicated, these would be non-trivial questions, never employing such bad tactics as multiple choice. The student mastering the material proceeds to the next unit. If not, there is further study and another test, until mastery.

Since in computer-based material this test is likely to be used repeatedly, the questions will not be individual questions, but will come out of question generators. With computer tests taken interactively, each 'question' can be a prescription for generating a question, and each student is very likely to see a unique question. This was easy to do in areas like mathematics and physics, but harder elsewhere.

We would never give the same test twice. The order, too, is random and can dependent on how the student performed on previous questions, tailored testing. The Educational Testing Service developed mathematical strategies for tailored testing.

In some situations, it may be necessary to develop a large bank of questions and pick randomly from them. But this is not as desirable, because it is more difficult to assure equal testing for each student who uses this material.

No single standard of mastery is required. One need not build into a system the idea that mastery represents 85%, 90% or 95%. The designers of the questions can make individual decisions of what constitutes mastery of the subject area.

As one gets to more complex cognitive goals, only very experienced designers, accustomed to the students with whom they are working, will be able to formulate these decisions, and they must be, we argue strongly, taken on an individual basis. There is no 'magic' from artificial intelligence or elsewhere that will guarantee what will happen with these decisions about mastery, at least with the current state of, artificial intelligence.

To continue to the next unit the student has to do almost perfectly on the unit test, thus assuring mastery of the topic. 'Almost perfectly' is again a designer decision.

10.3.1.2 Learning

Students study on their own or perhaps in small groups and choose their resources, different from the traditional classroom where the teacher controls everything. The student may have a variety of resources for learning. Many media may be involved. Conventional material such as textbooks is still important.

In the tests developed for beginning physics at Irvine, much of the learning material was part of the tests.

10.3.1.3 Structure

This organization for the Keller Plan, with and without computers, is in figure 2. We seldom see this interesting plan today.

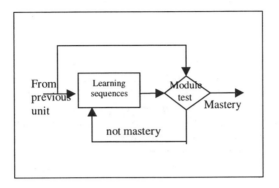

Figure 6. A Keller Plan unit

As we have hinted, replies to a student input to a question may contain learning aids. Thus, it might be desirable to tell a student immediately whether she; or he is answering the question correctly.

Furthermore, it is often possible to look for common errors. Experience shows that in many situations in science and mathematics there are only two or three common errors in working a particular problem, or perhaps a few more. Therefore, it is easy for the designers to recognize such difficulties in student replies and give specific help, direct useful comments about the error. This can be a very valuable learning experience.

10.3.2 Test-driven mastery plan course

The test-driven mastery plan is the next possibility. We used this structure in the physics course we developed at Irvine, in a timesharing environment, as already mentioned. The student sees the course primarily as a series of tests.

As in the Keller Plan, the material is divided into units, perhaps a weeks work for a typical student. Students may start a unit of work by taking the unit test, although this is quite contrary to what happens in usual classroom situations. During the test, the students may again have access to learning material such as text, lectures and built-in help. The help built into the quizzes can often locate student problems. In this situation much of the learning takes place during testing, a reason for calling this a test-driven approach.

Many times students will already know something, from previous background, or can quickly learn it through the test itself, given the assistance with student problems supplied there. In these situations it is a waste of student time to require learning situations first. Further, taking the test quickly shows the student just what is required in the unit.

It is possible to build the material in such a way that the decision on mastery is not entirely part of the program itself; the individual teacher can control it to some extent, if there is a teacher. So, teachers can adapt material to their own pedagogical preferences, although the program would have a default choice. This adoption would involve no programming on the part of the teacher. He or she would use an interactive program. However, as we educate more and more students, the concept of teacher is likely to vanish, particularly in distance learning environments.

10.3.3 Recognizing mastery

A variety of ways can determine mastery within the program.

A traditional way in a problem-oriented course, such as science and mathematics is by percentage of problems correct. But this does not allow for learning during the test, so this approach is seldom desirable.

Another approach is to keep score on correct problems, but to weight the last problems more heavily. Suppose that the subject, for example, is adding two two-digit numbers. The program can offer many such problems, perhaps offering some help after each problem. The designers might decide that having the last four problems correct would show mastery in the area of such problems. Or if this does not happen after a reasonable attempt, the program can move to a learning assistance program in the area.

Another strategy is for the designers to decide mastery at each point on an individual basis. There need not be a uniform approach for the entire program.

10.3.4 Combining learning and testing

In the mastery strategy emphasized in this book, the separation between learning and testing vanishes. They are intimately combined as part of the same learning material. Tutorial computer-based learning makes this possible, just as it was with human tutors.

Students do not know they are taking tests. We no longer have these stages, testing and learning, as unique processes. Learning and testing are part of the same process. Tests, in their usual form, no longer exist. We often discuss this desirable organizational method for future learning. The role of student evaluation is not to assign grades, but to assist students.

In another sense, tests are occurring every few seconds or minutes. Every question from the tutor, human or computer, is probing students to find where they are having problems, or leading them to discovery.

10.4 The Learning cycle

Robert Karplus, University of California, Berkeley, based on study of Jean Piaget's work, the learning cycle, suggested a method for organizing learning. . He used this approach in science units for young children. The learning cycle has three components: the experiential component, the formal learning component, and the testing component. These last two might be combined, as just suggested, so this is an extension of the ideas already presented about blending learning and evaluation. We consider this further at the end of this chapter.

10.4.1 Experiential learning

The experiential component of the learning cycle addresses two issues that are often weak or missing from learning. These are developing intuition and overcoming false beliefs.

10.4.1.1   Intuition

The first issue is the problem of developing intuition about the subject being learned. Intuition, intelligent guessing, is a vital aspect of learning today. It is often more important than other components of learning in everyday life.

An important part of this is developing a feeling of the appropriateness of a solution to a problem in advance of solving the problem so that one can determine if an answer is a reasonable one. This is an important skill for problem solving.

10.4.1.2   Student misconceptions

Experiential learning also has the important role of overcoming misconceptions, the naïve views in the area that the learner may have in advance of learning something. For example, in mechanics the student view of the 'causes' of motion is likely to be wrong, as seen by physics today.   Students tend to believe, from their everyday experiences on earth that a force must act to keep something moving, an Aristotelian view of motion rather than a Newtonian view. This problem would not be present for a student in space, with little gravity and friction.   Often the teachers themselves have the same misconceptions.

10.4.1.3   Tactics for experiential learning with computers

A major tactic used in experiential learning with computers is simulation.   A facility for exploring the subject area, gaining experience, is available on the computer.

It is important, however, to have more than just the simulation. A 'naked' simulation is common, but not suitable for the two roles just described.   We need simulations that behave in a way similar to apprentice learning. As the student uses the simulation, the program watches what is happening. It may offer suggestions about how to proceed, and it may check to find what the student has learned. The positive results of using the simulation are not left to chance.

The designers may also develop other tactics to meet these important goals. Students can confront evidence in conflict with their naive views, in simulations or in experiments. This is the Piagetian idea of accommodation. Advanced organizers may also help.

10.5 Complete mastery learning cycle,

The complete mastery learning cycle, the last organizational structure for learning we discuss in this chapter combines mastery in learning with experiential components. It is composed of computer simulation for experiential learning, mastery learning with mastery-based testing, and extensive student aid.

We can combine experiential learning with either of the master forms discussed so far, discrete testing or imbedded testing.

10.5.1 Discrete testing

This strategy combines the aspects, but still keeps testing and learning as separate ideas. Figure 3 indicates the structure.

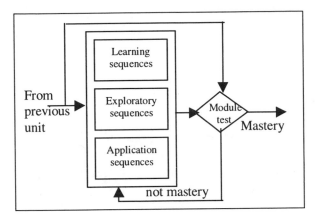

Figure 7. A complete mastery learning unit

10.5.2 Combined learning and testing

Our final approach, with the best of all these ideas, uses experiential learning with seamless learning. The student does not distinguish learning and testing. We have already given the details.

We strongly recommend this approach for the future.

Chapter 11

DEVELOPING TUTORIAL LEARNING UNITS

We must do more than bring computers into the classroom. Learning to mastery for everyone, at an affordable price, is the goal. To facilitate this, we must design effective development methods that are rooted in pedagogical theory and cognitive research. We must use our knowledge of learning, media preparation, assessment, computer science, and subject domain to bring about the change.

Independently of the issues discussed in this book, we badly need new units in all areas. This could be new textbooks; but we believe that an important direction to explore is the use of highly interactive multimedia tutorial modules for learning

11.1 History of the Irvine-Geneva system

In this chapter we present a system for development of computer-based tutorial learning units. It includes both methodology and software. We can imagine other systems that focus on highly interactive units.

This system, the Irvine-Geneva system, has a thirty-two year history. We did not begin at the University of California, Irvine, formulating the system in advance, or with developing software, but with trying to describe highly interactive learning material. The first such material was designed by Noah Sherman (University of Michigan) and Alfred Bork at the University of California, Irvine.

Students were to prove the law of conservation of energy in mechanics, beginning with the Newtonian Laws of Motion. They usually encounter such a proof in lecture or in the textbook. The program was highly interactive. The concept of the script, discussed further later in this chapter, evolved during this development as a way of representing the full design of this interactive program. The script produced for this program over thirty years ago still resembles the scripts produced today.

We implemented only enough software to turn this script into a running program. A student, David Robson, worked with Alfred Bork in developing this early software. This program led to our original testing with students.

This program, and other programs developed at Irvine such as Terra and Luna, designed by Arnold Arons (University of Washington) and Alfred Bork ran in a timesharing environment on a Sigma 7. We expanded the software as new demands arose from the designers.

The major project the Irvine group developed in this environment was a beginning quarter of introductory physics, mentioned already several times. Many competent designers and student programmers were involved in this development, and the ones that followed.

An essential change occurred at Irvine when we moved to from timesharing to an early personal computer, the Terak. Kenneth Bowles (University of California, San Diego) moved the portable Pascal to this machine, including the related operating system (the p-system), and we with Kenneth Bowles persuaded the vendor to add graphic facilities to an existing computer. We decided to implement a new software system in Pascal for dialogs developed for the Terak.

Stephen Franklin developed several critical pieces of software for this transition. The major such module was the ports system, allowing full control of the screen for both text and graphics. It would now be called a windows system, but screen control exceeded that normally available today for windows. Many of the facilities, such as student control of the rate of text output, followed the research on readability. The programs in the Scientific Reasoning Series, described in Chapter 7, ran in this environment.

Soon after this Bernard Levrat and Bertrand Ibrahim at the University of Geneva joined us in this activity, developing material with us. Bertrand Ibrahim was the leader in programming the online script system, a great advance in our capabilities discussed later.

We now discuss the four parts of the Irvine-Geneva system, management, design, implementation, and formative evaluation and improvement.

11.2 Management

The development of high quality learning material is a complex activity, with or without computer. Therefore, in all major developments we must pay careful attention to the management of the project. Unfortunately this is not always the case, particularly in university work where people have little training or background in management.

A development activity needs to keep on schedule, and to stay within cost boundaries, the function of management. We can take our lead from industry in this regard. Some computer management tools can help in this process, showing Pert charts of the activities, and providing scheduling reminders.

11.3 Design

The most important step in development of high quality learning material is design. Poor design can only lead to poor learning units. In the Irvine-Geneva system, we consider design in two stages, overall or total design, and detail design. We discuss these in the next two sections. Then we discuss recording the design.

11.3.1 Total Design

The task in this phase is to plan the general details of the learning units. We begin with information about previous learning material in this area, and with the experience and imagination of a group of teachers and other professionals.

This group is responsible for the total design. The details of design are the concern of the next phase. The results of total design will be a list of modules to develop, with a page or so describing each module.

11.3.1.1 Existing material and guidelines

Existing material in the subject area considered is important, including units with older presentation techniques. Much of our current learning material, mostly book- and lecture-based, is outmoded, from the standpoint of both content and pedagogy. The most widely used high school science and mathematics texts, for example, have changed little for many decades. Recent Project 2061 reviews of these texts for secondary schools in the United States have been dismal.

Many states, countries, and organizations also have guidelines and standards for learning. We examine these in the total design stage. For example, the National Council for the Teaching of Mathematics in the United States has such information, as do the benchmarks from Project 2061.

11.3.1.2 Brainstorming

The initial tactic for total design is brainstorming, generating quick ideas about the material to be developed. The usual rules of brainstorming apply.

- The group is encouraged to state spontaneous ideas.
- Ideas are written for the group to see, perhaps on a pad or board.
- No criticism or discussion occurs during brainstorming.
- Participants do not worry if their ideas overlap others.

There is a reliable literature about brainstorming, useful in looking for details.

The next stage is to organize these ideas. This usually leads to a list of modules. The final job of the group is to write the module descriptions, the input to the next stage of design.

11.3.2 Detail design

After total design, we still need all the details. The important task in this activity of detail design is to generate all the particulars of the design, beginning with the data from total design. This information will be the input to the implementation activities.

11.3.2.1 Design groups

We typically work with design groups of about four. These are usually excellent teachers in the area being developed An international group is desirable, as the countries involved, with different cultures, may have different insights into learning.

The design process tries to capitalize on these teachers' experience and contacts with students on an individual basis, their insights about student problems and their ideas about how to help with these problems.

> To generate these kinds of adaptive lessons, authors or other members of the development team need extensive experience teaching the given subject matter to the target population on a one-to-one basis. Developers need to know the kinds of misunderstandings that often occur, so that they can generate diagnostic questions and provide suitable feedback. This kind of information cannot be obtained from instruction by lectures. It is gleamed from interactions with individual students and supplemented from the results of trial lessons when they are under development.
>
> Esther Steinberg
> Computer-Assisted Instruction
> Erlbaum, NJ, 1991.

It is useful but not essential to have someone in the group who has previously designed interactive material. As more material is designed, this will become easier to do. We tell the designers not to worry about implementation, but to design the best possible material for learning. We tell them that we can implement anything they design! Very occasionally this produces a sequence more suitable for companies such as DreamWorks, than for us, but this is very rare.

A design group at Irvine normally works together for a week, full time. In a large project several groups may work at the same time, getting together during meals and breaks. In a week, the group will typically design about two student hours of highly interactive tutorial learning material.

The design activity should take place in a pleasant relaxing environment away from the usual locations of the participants. We do not want the designers to be involved with their usual activities, but rather to think and work primarily on this design during the week.

11.3.2.2 Getting started

The first activity of the design week is to prepare the designers for the task ahead. This is typically the task of the first morning. We begin with the notions of highly interactive tutorial learning, as described in this book, including the notion of the design script, discussed soon. This approach will be new for many of the teachers in design groups. This will include an understanding of the zone of proximal development.

The second task is helping the designers to understand the dynamics of working in a group. Many teachers have had little experience with such groups. They need to realize that everyone in the group is equal, that there are no leaders. They must also understand the possibilities of conflict and the necessity of quickly reaching an agreement on disputed issues.

In the future, designers will use tutorial computer-based learning materials before coming to the design sessions. We plan to develop such material.

11.3.2.3   Tasks for the designers

The designers make ALL the decisions about what is to happen during the learning activities. We emphasize that the program does not make these decisions, except as instructed by the designers! The main purpose of this design activity is to find student learning problems and to help with them. The process continues until each student reaches mastery.

The teachers design the messages the students will see or hear, or both, mostly questions to the students. They ensure that the interactions are frequent and of high quality. They describe how to analysize student input, either by typing or by voice. They show what to do in each case, indicating the flow of the program.

Another type of decision the designers make concerns saved data. They decide what to store, and how to use the stored information, to assist learning. There are two purposes for this saved student information, guiding the future learning of the student and helping with formative and summative evaluation.

We have previously discussed how this process of questions, responses, analysis, and stored information shows what learning material to present next to the student, keeping learning in the student's zone of proximal development.

The designers also specify all the media to be used, pictures, video, and sound. These will typically be verbal descriptions, perhaps with sketches. In some cases, as with the Japanese material described in Chapter 7, video is already prepared and available to the designers.

11.3.2.4 Review of the detail design

We review the design before the implementation stage. This requires people who are already familiar with the script strategy, described next, perhaps because they have been involved in other design projects. They may be members of the project staff, or they may be located remotely.

This review may determine weak parts in the design, where material requires strengthening for best pedagogical results. A problem we see, with designers accustomed to writing text, are long verbal descriptions. Another is the use of extensive computer animation with little learning value. In both cases, interaction is weak.

11.3.3 Scripts

The designers must have a way of recording all these decisions. This leads us to discuss the script, the document that records all these design decisions. The script is the input to implementation, the next stage in development.

Initially scripts were on paper, usually large sheets so that the designer can see much that happens. As mentioned, Bertrand Ibrahim at the University of Geneva developed the online script. With the script editor, designers can enter the script at the computer. The choice of paper or computer can now depend on the design group.

11.3.3.1 The online script

The computer-stored script has major advantages for the designers. In the complicated logic of a script, the designers can neglect some details that need to be completed. The software can locate such errors for the designers.

The script can also be modified in any way in its computer form, as changes are made by the designers and on the basis of formative evaluation. Previous versions of the script can be stored as backup. Editors can examine the texts and make changes where necessary.

11.3.3.2 Script elements and examples

The following examples are screens seen in using the script editor. Note that the full text, for example in the message 'We have a' in the following script, the full message is not shown on the screen. This is to allow more of the logic to appear at one time on the screen. Designers can always ask to see the full message.

In the first example, Bertrand Ibrahim has labeled the types of nodes found in the script. These labels are not part of the script, but the designers can identify them from the visual information.

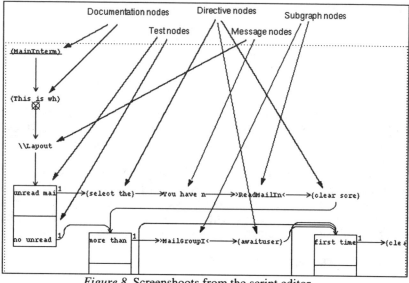

Figure 8. Screenshoots from the script editor

Much of the script concerns testing, in the boxes, essential in highly interactive tutorial material. This compares to the slightly interactive learning in much current online software. This script, with a large number of test boxes, does not resemble a book or a lecture. Tutorial learning is very different than learning with information transfer. The lines with arrows show the flow of the program.

Many of the tests refer to what the student has done before. Thus, this may be the first or the second time that the student has been at a given point. Directives, comments, are mostly messages to the programmer about what should be happening.

We show next a second script fragment. Unfortunately, you may not be able to read all of this, as the screen is much larger than the space in this book. Nevertheless, the notion of the amount of testing needed, and the many paths possible in the logic, are again apparent from the large numbers of rectangular test nodes.

By calling these fragments, we emphasize that desing with the script editor, is modular and we end up with many small scripts rather than one big monolithic script. We only see in these pictures a very small portion. The scroll bars in this script allow the designers to move easily to other parts of the script.

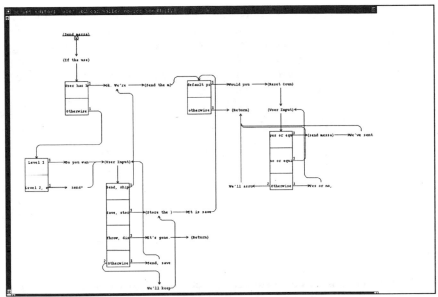

Figure 9. Second screenshoot from the script editor

11.3.3.3 Language support in the online script

One advantage of the collaboration between a United States group and a Swiss group was that we addressed the problem of multiple languages early in the project. So multilanguage support is an integral part of the Irvine-Geneva system. This is very important for learning units used in many languages and cultures, as suggested in this book.

The script, seen as a visual object by the user, is stored internally as collections of two files. One in each pair contains all the logic of the program, and the other contains the messages and other elements that are dependent on language. A new language uses the same logic file.

A teacher, who knows both languages, and how students use the old and new languages, is the best translator. Cultural differences are also important at this time, although these may have been partially addressed by having designers from many cultures involved.

The script editor includes a version control mechanism which makes it easy to see what changes have been made since the last implementation. It also includes translation management tool that deals with various translations, including a version-control mechanism in order to see what changes need to be made to some language versions after an update in one of the languages. This is a major bookkeeping problem as a given segment is available in many languages.

11.4 Implementation

This stage of development moves from the reviewed design, a script, to a running program. The major activities are programming and media preparation.

11.4.1 Screen design

The layout of the screen affects how easy it is for students to learn. A professional designer does this screen design in the Irvine-Geneva system, ideally one with strong interests in learning. Screen design depends on the complete script, determining what should go on the screen.

The often ignored research on readability is valuable in screen design as mentioned. Online learning units are often particularly bad in this regard, with text-dense screens and useless pictures.

The following ideas guide this design, supported by the ports software.

- Only material currently pertinent should be on the screen. Other text and pictures may distract the learner.
- The screen should have little on it at a given time. Blank space is free on computer screens, unlike the situation with books.
- Short lines, not right justified, improve readability.

- Lines should break at natural phrase breaks. That is, a phrase like 'in the house' should appear on one line, unlike it appears here.
- No hyphenation.
- Separate screen windows should indicate different purposes. This should be consistent throughout the program. For example, one window might be for questions to the student, another for student replies, and a third for help to the student.
- Students should be able to control the rate of text display, at any time.

We have partially designed an interactive program for screen designers, based on these ideas. The program will use the scripts, showing what material is to be on the screen. The screen designers will not need to do any programming.

11.4.2 Program

We are creating a computer program in this stage, based entirely on the script prepared and reviewed in the design stage. In the time before the script editor, student programmers using a full standard programming language created this program. We used Pascal initially, as mentioned.

The script editor now allows us to produce much of the program automatically from the stored script. This can be in Java or in any language specified by the project. This might imply a version for computer A and another version in a different language, perhaps, for computer B; so a program for any computer can be created from the design. The script is to a great extent hardware and software independent. If a new computer appears, with new languages, the script can cope with this change. Later we will consider the possibility of interpreting the script when it is running, preparing a new interpreter for each language and computer. The speed of contemporary computers will allow this approach.

Although we can look forward to having most of the code generated by the computer, humans must still write some of the program. The designer can request anything, so we cannot anticipate all possibilities in automatic programming from the script. The parts written by people must fit with the computer-generated code. The script editor helps with this integration process.

11.4.3 Media

In addition to programming, media (photography, computer graphics, video, and sound, for example) must be prepared to meet the demands of the designers. This is a job for professionals in the area. Additional design may be needed before production, as the designers are not experts in the field and so their verbal descriptions of what is needed may not be adequate for production.

11.4.4 Integration

The parts, programs and media, must all be put together now. This produces the first version of a module that has all the ingredients, an exciting moment in the project. However, there are, in this complex activity, likely to be many problems.

11.4.5 Beta testing

After a program is available, we can begin testing to find these problems with how it runs, bugs. Beta testing is standard for the computer industry. It is not evaluation, not common with software. The aim of this testing is to produce a stable program suitable for evaluation.

The initial testing can occur internally to the project. This will uncover the major difficulties. Then beta testing can involve sizable numbers of typical students. This continues until the program is stable, running without problems.

11.5 Formative evaluation and Improvement

We now have a stable program. The next step in the development process is second only to design in its importance in leading to effective learning materials. No matter how competent the designers are, they cannot anticipate all possible situations with students! So extensive testing with large number of typical students, and improvement, is indispensable before general release of the learning modules, the function of formative evaluation.

It is vital to involve all the types of students who will be using the final product. If the product is for international use, students from various countries, perhaps using different language, should be involved in evaluation.

Professional evaluators will guide this effort. Our goals suggest what we look for in gathering the data for formative evaluation.

- Do students understand the language used at each point in the program?
- Does the program do an adequate job of analyzing student input?
- Are there problems with voice input?
- Do students stay in the program for long periods, even in environments where they are free to leave at any time?
- Do students learn in a reasonable amount of time?
- Do all students achieve mastery?

11.5.1 Gathering information

We gather information about student performance in several ways. The major method will be through computer-saved data, while students are running the program. We need to gather some data about each student, perhaps as part of the learning activity. The designers specify much of the information to be stored.

We also store all student responses not anticipated in the program, to see if we are missing something. When the computer is in free environments, we can record the places where the student leaves the program, to look for points that are weak motivationally. Information about how quickly each type of student achieves mastery at each point will also point to places needing additional work.

In addition, evaluators, people, gather information. All information is on the computer. Since large numbers of students will be involved, the student database will be extensive. Programs will assist in arranging this data for the improvement stage that follows.

11.5.2 Improvement

The data can lead to substantial improvements in the program. The staff of the project can make many of these changes. Some changes are directly in the stored script; new automatic code can be generated. We might require new programming and media preparation.

If some sections prove to be weak, additional design sessions may be necessary.

A new improved program will be the outcome of this step.

11.5.3 Another cycle or two

At least one additional stage of formative evaluation and improvement, or perhaps more, is highly desirable, using the improved program. The details are the same as those suggested. Finally we emphasize again the critical role of evaluation and improvement in leading to effective tutorial learning modules.

11.6 Summative evaluation

An important aspect of considering learning material is how one set of material for learning a given topic compares with another. This comparison can have many dimensions: Enjoyment of learning, time to learn, learning to mastery, retention of learning. The comparison materials might be standard material (such as a lecture-based course), special material, or other computer-based units. We insist on data, not opinions.

Such evaluation should have a strong component of longitudinal evaluation, where we watch the students for many years. This requires that we have continuing funding for this activity, and that large numbers of students are involved, as we will not be able to find all of the original students.

Longitudinal evaluation is not easy, but it is essential. The long-range effects of curriculum materials are the most important aspects.

11.7 Improving the Irvine-Geneva system

The system outlined in this chapter has been under development for over thirty years. We believe it is well suited to the task of generating tutorial units. However, improvement is possible and desirable. We have already mentioned some of these ways, such as an interactive facility for the screen designer. The initial interactive units for helping designers understand tutorial learning, is for use before the design sessions.

We can increase the percentage of the program written by the computer from the script, perhaps reaching ninety percent. We can add facilities for suggesting points in the script needing improved interactivity. In addition, many other improvements are possible, given the resources for this work.

We finally emphasize again that other systems might generate tutorial learning materials.

Chapter 12

COSTS OF HIGHLY INTERACTIVE LEARNING

One of our important goals in Chapter 2 is that learning should be affordable for all, not just in the wealthy parts of the world, but everywhere. This is critical for worldwide learning. In this chapter, we discuss these issues of cost, comparing costs to some extent of different types of learning, although the data is often missing. Nevertheless, as the other visions indicate, we must also be concerned with the quality of learning. We want costs for learning that support all of our goals.

Computer-based tutorial learning, as discussed so far in this book, may provide such affordable learning. But we need more experience with it to be sure. At least we can say that tutorial learning with computers is highly promising, both from a learning viewpoint and a cost viewpoint. Therefore, this attempt is incomplete, with many uncertainties.

It is clear to us that none of the strategies for learning now receiving large attention will be economical for worldwide learning. Most of them are not scalable to the numbers of learners needed to achieve our visions.

Even in the United States and other developed countries the current approaches have little possibility for attaining the funding they need to be fully workable, and effective new learning approaches at lower costs are badly needed. That is one reason why we are seeing the rising participation on for-profit companies in the market.

12.1 The economics of learning

The existing literature about the economics of learning is not impressive, and often does not look at all the costs involved. Often important factors are ignored. Even the question of what is most important in calculating costs is often fuzzy.

The bottom line we argue is what it costs to provide *a student hour of vgry high quality learning material to students, anywhere in the world.* We need a system in which the cost per student declines or perhaps stays the same as the number of students increases, an approach scalable to very large numbers. This chapter is based on that factor.

We now discuss the cost factors for highly interactive tutorial computer-based learning, as stressed in this book. A number of components go into determining these costs.

- Development costs
- Marketing costs
- Administrative costs
- Delivery costs
- Profit, if material is distributed commercially
- Costs for usage with poor individuals, including other languages and delivery
- Funds for additional development and distribution

Each of these items will be discussed separately for computer based tutorial learning. All are important in examining cost for a student hour. Some are easier to consider than others are. Some, such as delivery costs, depend heavily on the numbers of students involved. Others are more difficult to consider because of the lack of current experience, so that some reasonable speculation is necessary, later modified by practical experience.

After considering each of these components separately, we will return to the critical problem of the cost of a student hour. We would like to compare costs for tutorial computer-based learning with existing costs for learning, but the data for this is often not available.

Cost is not the only factor in the vision in Chapter 2. Learning must also be effective for all students. That consideration is vital in this book.

12.2 Development costs

This is the cost factor most often discussed, often poorly. The central consideration, usually ignored, is that development costs are connected to some extent with the quality of the learning product. Therefore, this issue cannot be discussed independent of quality, although quality is difficult to determine without extensive experimentation.

High developmental costs will not guarantee excellent learning materials, but earlier work indicates that low costs will almost always lead to poor learning modules. Our interest is in very high quality learning, so that the learning will be effective for a wide variety of learners.

12.2.1 Difficulties in determining development costs

In many universities a course costs $2.41 to develop (we exaggerate slightly, but it is not clear that this is high or low!) Often the figures for development given ignore such factors as overhead, management, faculty salaries, and evaluation. No careful data is maintained.

One of the worst examples of this difficulty in determining costs of development is with computer-based learning. Often figures such as 300 faculty hours to develop a student hour of material are quoted, clearly we believe not enough for high quality individualized material, as we will see shortly. Those who give such figures seldom look at the full costs, including the items beyond development.

Even professionals tend to underestimate costs. An experience of one of us (Alfred Bork) illustrates this. He was sitting in the fancy offices of a prominent and successful commercial source for computer based learning units, comparing costs for development. Those quoted by the company seemed low. On discussion, it came out that the costs the president was giving did not include some fraction of the costs of that fancy office, or of secretaries or other workers. Further, it did not include any expenses for evaluation of the learning material.

Our experiences with book publishers, although not recent, show a similar problem of not knowing full costs. This seems strange in a commercial environment, but it does happen. Producing a new major textbook is expensive, it is clear. An example, not in education, in the early days of personal computers of not knowing costs of production was a price war between two companies that ended when one company discovered that it was selling computers at less than cost. As a friend commented, this is a situation where volume does not help!

12.2.2 Curriculum development costs without computers

There is a history concerning the costs of developing high quality learning units, some quite old. Two examples give us benchmarks, the post Sputnik development in the United States, and the United Kingdom Open University. In neither case were tutorial computer-based learning courses developed, but they show the expenses of quality curriculum development.

The United States launched a major curriculum effort after the USSR placed Sputnik in orbit in 1958, described further in Chapter 13. The Open University, mentioned many times in this book, began in 1968. A hallmark of its work was and is a careful development process. Large teams of specialists are involved for several years.

In both cases computers were not heavily involved, except in recent courses developed in the Open University such as the foundation course in computer science. No highly interactive tutorial units were developed, as mentioned.

The costs for quality development were similar in these projects. A year of student work required millions of dollars to produce. In today's dollars that would be more, due to inflation. This gives us a benchmark for the system proposed.

12.2.3 Costs for tutorial computer-based learning units

The group at the University of California, Irvine, has been developing tutorial learning modules on computers for over thirty-two years, since 1968. So we have good data about costs of such development. The expenses we encounter are similar to those in the two examples just mentioned, even though these previous programs were not primarily developing material for the computer. Quality curriculum development, whatever media are involved, is expensive. Much of the development costs are in pedagogical design.

As discussed in Chapter 11 we consider development to have four stages, management, design, implementation, and formative evaluation and improvement. The budget will not be the same for all projects. For example, if extensive high quality video work is needed, implementation of this project will cost more than other projects not requiring this.

12.2.4 Example of a budget

We are often asked about how we determine the developmental budget for a project. To illustrate this in detail, the budget for an unfunded project for about a year of student work is given here. As stressed this could be different in another project. A reader who is not interested in such detail can move rapidly through the budget, or skip it entirely.

12.2.4.1 Management

Table 1. Management cost

Management cost	Year 1	Year 2	Year 3
Principal Investigator 1/2 time	58500	59670	60830
Co-principal Investigator 1/4 time	23400	23868	24345
Project Manager	61500	63653	65880
Graduate students 3	57269	58414	59583
Graduate tuition (3)	45473	48684	52128
Administrative Assistants	37505	38818	40176
Supplies	5000	5000	5000
Copy and phone	5000	5000	5000
Computers - workstations	6000	6000	3000
Travel - domestic	4000	5000	6000
Travel - foreign	7000	9000	9000

12.2.4.2 Design

Table 2. Design cost

Design cost	Year 1	Year 2	Year 3
	60 hours	60 hours	10 hours
Travel for designers	144000	144000	24000
Subsistence for designers	144000	144000	24000
Stipend for designers	90000	90000	15000

12.2.4.3 Implementation

Table 3. Implementaion cost

Implementation cost	Year 1	Year 2	Year 3
Programmer	55350	57287	
Programmer - ½ time			32940
Student programmers	61858	42682	22088
Graphic consultant – screen	5000	5000	
Media production Designer - ¼ time	10000	10000	
Media preparation	40000	30000	10000
Student assistants	21042	16332	

12.2.4.4 Formative evaluation

Table 4. Cost of Formative evaluation

Formative evaluation	Year 1	Year 2	Year 3
Consultants	20000	10000	
Student assistants		54440	22540
Equipment – informal use			
Kiosks		15000	10000
Computers		40000	10000
Software		5000	2500
Shipping equipment, kiosks			10000
Improvement Programmer - ½ time		31826	32940

Formative evaluation	Year 1	Year 2	Year 3
Student programmers			22088
Additional design sessions			30000

12.2.4.5   Advisory group

Table 5. Cost of Advisory group

Cost of Advisory group	Year 1	Year 2	Year 3
Travel	8000	8000	8000
Subsistence	8000	8000	8000
Consultant Fees	10000	10000	10000

12.2.4.6   Special activities

Table 6. Cost of special activities

Cost of special activities	Year 1	Year 2	Year 3
Improvement of production system			
Design	20000		
Implementation	10000		
Testing	3000		
Consultants	7500		
Project Web site	10000	5000	5000
Scientific Reasoning Series	0	0	0
Voice input	5000	0	0
Better graphics	10000	0	0
Web enabling			0
		10000	
		0	0
Final report		0	10000

These figures include benefits, but they do not include the costs of space for the developmental activities.

12.2.5 Summary of development costs

We can summarize this discussion of development costs by saying that about $30.000 is needed to develop a student hour of learning material, understanding that a particular project might cost more or less than this.   This is divided between management, design, implementation, and formative evaluation.

Calculation of the cost per hour for student usage for just the development costs is easy to do. We need only divide the cost to produce an hour of learning material by the number of students who will use the material.

Thus if the costs are $30,000 to develop an hour of material, and 30 students use the material, the development cost is $1000 for each student hour, very expensive for most development, but competitive in some environments. If 10,000 students use the material, as in some of the Open University foundation courses, the cost for developing a student hour is $3.00. If a million people use the unit, as they might in a literacy program as in the sample budget – the reader can do the arithmetic.

The lesson here has been long understood at the Open University and in the other mega-universities, but sometimes little appreciated in the United States. Expensive high quality development can lead to very low student costs if the material is of sufficient quality to be very widely used, and it is marketed effectively for such extensive use.

For worldwide mass education, our surprising conclusion is that the cost of development is nothing, even for expensive development, if viewed from the standpoint of each student. To be used with a very wide audience, the learning material must be of high enough quality to adapt to the needs of each student, which implies expensive development such as that discussed in this book.   There is no substitute for learning quality!

Worldwide use also suggests we must allow for costs of adapting tutorial modules to new languages and cultures. Although we do not have sufficient experience in such conversions, we estimate from our efforts that the costs for each new language will be about $3.000 for a student hour, one tenth of the initial development costs. Again, this figure would vary from culture to culture.   This assumes a developmental system, as in Chapter 11, allowing many languages.

12.3 Marketing costs

The history of marketing computer learning material is not impressive. Further, costs are difficult to estimate, given our lack of experience. We need to explore a variety of marking approaches, including making early versions or subsets widely available for very little cost, perhaps free.

12.3.1 Interesting learning modules

An important marketing consideration is that the materials, as stressed earlier, must be intrinsically motivating. If students enjoy using them, this will encourage further use. We cannot simply advertise that the units are 'fun' to use, as often happens now. Rather, as already indicated in our discussion of development procedures in Chapter 11, we must verify empirically that motivation is effective in all parts of the program.

We stress again that this does not mean that the programs should use such gimmicks as puppets, loud music, and similar ideas gathered from the film world. Highly interactive materials, with frequent meaningful interactions, are, we find, intrinsically motivating.

12.3.2 Evaluation data

Another important factor in marketing will be the comparisons of these units with other ways of learning the same material, in terms of ease of learning, time for learning, retention, costs, and other factors. If we can show clearly and forcefully in summative evaluation that these materials are highly superior in these ways to conventional learning, as discussed in Chapter 11, marketing will be easier. We believe this will be possible for computer-based tutorial units.

12.3.3 Marketing in developing areas

In the areas of the world with little learning today, both developing countries and poor areas of developed countries, marketing will be relatively easy. There little quality learning is available, although there is a growing realization that learning furnishes the best path out of poverty. However, funds may not be available, an issue to be discussed later.

12.3.4 Costs for marketing

The question of marketing costs is one where we need better data from experience. So the following considerations must be regarded as tentative. A full model of costs in this area is likely to vary from area to area.

We expect, as with delivery costs, that the costs will increase with the numbers of students, but less than linearly. In our tentative cost model, we will assume that total costs for marketing go up as the third root of the numbers of students. So the cost per student goes down by the 2/3 power of the number of students. Experience can improve this model.

12.4 Administrative costs

A major cost of learning today is the administration of the learning establishments, schools, universities, training facilities, educational agencies of the governments, standards and accrediting organizations, and many other such facilities

With tutorial distance learning the cost factors for administration will be very different. The learning modules will reside on computers, perhaps in many parts of the world. Student records, gathered frequently as the students use the modules, will also reside on (probably different) computers. Communication with users, discussed further under delivery costs, will be another cost at least partially administrative. Information about available units must also be provided. Credit, degrees, and certification is another consideration.

Administrative costs also include planning for new development, and maintenance of existing learning units.

We expect usage to suggest points in the material that can be improved, an important factor in maintenance. These might be changes in the logic, made in the online script, changes in text, and changes needed because of new forms of hardware.

When material is available in 44 languages, for example, the problem of changes in text can be a major one! If the French version has a significant textual change, suggested by evaluation and usage, we must propagate this change to the 43 other languages. The Irvine-Geneva production software already discussed helps with this problem. As new types of hardware appear, we might want to propagate learning units to these new machines, particularly if there are economic gains in doing this.

We assume for administration a cost model like that for marketing. A separate factor given later accounts for multiple languages.

12.5 Delivery Costs

Most of the expenses of traditional school, university, and training education come with the delivery costs. One must build buildings, furnish them, heat and cool them, maintain them, garden around them, and provide facilities for and pay teachers and professors. All of these are expensive activities, and none are needed for the type of distance learning described in this book. We do not know of any recent studies indicating the cost per student hour for traditional learning. One would think that at least public universities would be obligated to make such studies.

The Open University also has high delivery costs, although much less than those of traditional universities. Much of this cost comes from supporting the human tutors that offer individualized attention.

There are many options to consider in determining the cost of delivering computer-based tutorial distance learning. And new developments in technology may influence these costs. As the amount of learning delivered this way grows greatly, we will need to reconsider frequently delivery methods, looking for the least expensive approaches at a given time for delivering highly effective learning. Note that we say 'approaches,' a plural word.

12.5.1 Method of delivery

As suggested in Chapter 8 and elsewhere, a variety of delivery methods should be used, to secure the greatest possible coverage. The mixture of methods will change in time. Initially we suggest both Internet and CD-ROM (or DVD ROM). The materials will appear to be identical to the students, regardless of the delivery method.

Eventually the most likely delivery method is likely to be by satellite. The student's location is then not a factor; learning for everything can be available everywhere. Two way satellite connections are already available in some areas from StarBand, and other companies, such as Hughes, will soon enter the market.

A recent study done for the World Bank shows that for relatively small numbers of students, in the hundreds, satellites are the cheapest delivery method. This study was based on a two-stage satellite system, one satellite for international distribution and regional satellites for local distribution. Phone lines and cables will not be needed.

As with marketing and administration, costs for delivery will increase with the numbers of students, but not linearly. We assume that the factor needed is the square root of the number of students.

12.5.2 Computers for student use

Another delivery cost component is the cost of the computers for students. Electricity will be an option, as solar powered laptops have already been build.

While the learning units will run on ordinary personal computers, and probably will in developed areas, they can work with much cheaper learning appliances, costing perhaps one hundred dollars. If such a unit will supply on average 5000 hours of learning before breakdown, a reasonable figure, the cost per student hour for this component of delivery is two cents. Therefore, this is a minor component in costs when we consider large numbers.

12.6 Profit

The units considered may or may not be considered as a source of profit. If a nonprofit organization, such as a government or the World Bank, funds them this factor can be ignored. But it seems likely that, in some cases, organizations seeking a profit may want to enter this market. We already see examples of this. This material could be a very large source of profit in the future. Companies will determine individually what profits to try to attain, depending of the market, so the figure given below might vary, as will many of the factors here.

As we will mention, units may be available for profit in some regions, and nonprofit in others.

This factor and the next two have a different economic basis than those just discussed. In each case it seems best to regard each of them as increasing the cost of a student hour by a fixed factor, a percentage increase. We suggest for all three items an initial percentage increase of twenty percent. This number will need to be adjusted as more experience is gained. It is lower than companies normally seek. But since the market may be very large, and there may be competitive companies, it may be possible.

We will note additional details in the next two items.

12.7 Costs for usage by poor students

One very intriguing possibility we have stressed is that sales of these materials in developed areas could furnish part or all of the funds for language and cultural conversion and distribution for the rest of the world.

We need to explore this idea in more detail. The middle class and wealthy of the world, while gaining superior learning for their children, will provide for learning in areas in poverty. There is a poetic justice in this!

12.8 Funds for additional development and distribution

Development is not a one-time activity. Courses must be revised, and additional material developed. Again, the United Kingdom Open University has considered this.

Some new development would come from private sources, or international organizations. In some areas of learning these sources might not be sufficient. We can plan for such development by adding appropriate costs to our budget. So current sales would support new development.

12.9 Complete costs for a student hour

The surprising conclusion to this chapter is that some of the items that might have been thought of as expensive turn out to be of almost zero cost in their contribution to an hour of student learning. Careful development is such an item, when large numbers are considered. Current attempts to save money on development might better go into developing more effective materials. The cost of the computers for students is another such item, if the computer is fully used.

We now develop a model for costs. This discussion indicates that we should consider the following variables in determining the cost for a student hour of highly interactive tutorial material

- N – number of students
- D – cost to develop an hour of material
- L – development for another language, as fraction of developmental costs
- F – number of languages
- S – Factor for materials distribution
- C – cost of a student computer

- H – hours of student use in computer lifetime
- M – Marketing basis
- A – Administrative basis
- P -- Factor for profit, poor students, and additional work

Our formula for the cost per student hour for excellent tutorial computer based learning units delivered at any location is the following.

$$P(D/N + (D/N) L F + S/N^{1/3} + M/N^{1/2} + A/N^{1/2} + C/H)$$

The first term, D/N is the cost for development, the second the cost for moving the units to other languages, the third the cost for distribution, the forth is marketing costs, the fifth is for administration and the last is the cost for student computers. The multiplicative factor P allows for profits.

Some of these components are relatively unknown at present. Nevertheless, this model leads to a modest range of at most several dollars per student, for large numbers of students.

12.10 Other forms of learning

To understand this data, we need to compare it with other forms of education. This is often difficult to do, as already noted, because much available information does not give the complete costs.

12.10.1 United States K-12 schools

Our information about schools comes from Frank Withrow, in the United States Department of Education for many years. He says in a personal communication, "In K-12 you can use a range of from $7 to $10 per hour per student. In the richest districts it can be as high as $20 and in the poorest districts as low a $2. One of the greatest difficulties in figuring this type of cost is to amortize building, equipment and facilities. . . .. The average life of a building should be about fifty years for cost considerations. So a $50 million high school that serves a 1,000 students would add a cost of about one dollar per hour. "

Note that even in the United States we expect to deliver superior learning at a lower cost with the new approach.

12.10.2 University costs

We do not have data for university costs. The United Kingdom Open University has costs of about half that of traditional universities.

12.11 Up front costs and continuing costs

A final financial note is that different systems require the money at different points in the learning cycle. Much of the costs of traditional learning are continuing costs in teacher and administrative salary. Here, buildings are an advance cost.

However, in tutorial computer-based learning we need funds for development in advance. A preliminary investigation suggests that we will not show a positive cash flow from this activity, pursued on a broad scale, for four or five years.

Chapter 13

INITIAL TUTORIAL LEARNING UNITS

Where do we start? That is, what initial development efforts should we undertake? We need to consider this for both the initial experiment, and for full development, both described in the next chapter. The possibilities are different for different levels of learning. This is far from an exhaustive list.

13.1 Pre-school and early learning

The great importance of early education is now widely recognized. Very young children are already wonderful learners, and what happens at this stage affects much of later education. The promise in this area is very great, both from a learning standpoint and from a marketing standpoint.

Although there is much Web and CD-ROM material for preschool use, most of this is poor material, with weak levels of interaction. The material available is not tutorial. Often these materials offer only entertainment or information, not learning.

The market is already large, and potentially very large, both because of numbers involved, because learning is often not available at this age, and because many parents are eager to provide education for children at this level

There is much experience in what young children should be learning. We can draw on the experiences of the Montessori programs, the headstart programs, and other successful programs.

13.1.1 Early learning of reading and writing

The fictional accounts of learning we reviewed all stressed the importance of early reading. We need to design and evaluate computer-based tutorial learning units that pick up just after language is learned, so that the spoken language quickly leads to reading and writing. Just as almost everyone learns to speak, almost everyone should learn to read and write. There is no reason for illiteracy in today's world! We expect voice input to pay an important role in this very early learning.

13.1.2 Living with others

Another important aspect of this very early informal tutorial learning will be to prepare people to get along with others. Nonviolent ways of viewing the world should be learned before five years old. We want happy individuals who respect the rights of all others. The only possibility for this to happen is with tutorial computer-based learning units. There are too many people for other approaches.

13.2 Home Learning

The market for home learning is not limited to preschool levels. It can be important at any level. This rapidly growing market legal in many areas could be much larger.

Home schooling is most effective when a knowledgeable and competent mother (very occasionally a father) can devote large amounts of time, serving effectively as a human tutor. So far, the numbers of students learning this way are small; often private agendas outside of learning motivate this activity. . Few parents in the world can devote such large amounts of time to their children. In many cases, there are NO parents to play this role, or the parents are far too busy. Standardized tests increase the problems.

The curriculum units available now for home schooling are mostly those available in the schools, so information transfer is the paradigm for most home learning. The 'tutor' changes this. Computer-based tutorial material would greatly expand the possibilities. The home could be a major center for all learning. However, not all students have a home that could support such an activity, particularly with poor families worldwide.

The home market may be important in moving us toward distance learning, as a supplement to schools or as a partial replacement for schools. Many parents are unhappy with their children's schools. They try to aid school learning with home learning. There is already a market for computers in this activity. Preschool learning as mentioned is one such large home market for highly effective carefully evaluated learning units.

13.3 Training

This is a large expensive activity, mostly conducted at present by companies. It seldom uses interactive tutorial approaches.

A direction that would furnish a good beginning point would be generic training needed for many companies. An example would be material based on national or international safety procedures. This could be widely marketed.

In addition to the traditional training programs in companies, we now are seeing many examples of corporate universities. They activities are frequently aligned with older traditional universities, particularly for the degree and certification purposes. Tutorial learning would help greatly.

13.4 Adult education

This is another important area with little systematic development. As our population ages globally, it assumes increasing importance. Marketing prospects are good.

Such places as the public libraries now are important centers for adult learning. Shopping centers, museums, and similar informal locations are likely sites.

Studies could investigate what materials should receive priority. The market could be very large. Perhaps the best place to start would be with all the illiteracies, already mentioned, such as that in science. This is a problem for adults as well as children.

> Learning can no longer be viewed as a ritual that we engage in during only the early part of one's life.

UNESCO
Learning Without Frontiers
December 1999
http://www.unesco.org/education/lwf

13.5 Schools

Schooling is potentially a huge learning market, one of the largest. However, it is also a difficult market to penetrate, given the large forces in the developed countries that depend on the current situation and that would resist change.   'Improvements' are frequently suggested, by politicians and others, but few lead to better schools.

Educating young children in schools involves far more students than higher education and lifelong education, although this may change, for adult education. This is currently the largest market, so the one with the most potential profit. Much of the costs for schools are currently for teachers and administrators. It is now a labor-intensive process, based primarily on information transfer as the learning paradigm.

However, the strong forces existing in the school establishment will not be easy to change.   The only way to move to the tutorial approach will involve major changes.

13.5.1 School problems

Unlike universities, schools have long been subject to extensive criticism. Perhaps the most famous report on this topic in the United States, not the first or the last, is *A Nation at Risk: An Imperative for Educational Reform,* published in 1983.   The title expresses the negative view of our schools and the possible unfortunate future impact on the country.   Many such studies exist.   Nolan Estes described our schools as 'a cage for every age.'

In many parts of the world schools do not exist, and have little prospect of meeting the demands for learning. Hundreds of millions attend no schools, and many have only a minimal education.   About one billion adults are illiterate, the majority girls.   Later we return to this important issue of international education. Progress may be easier in those areas, as there is little now and the need is very great.

These reports on school problems and the literature that followed have suggested a very wide range of cures, mostly for developed countries.   These include national goals, curriculum frameworks, higher standards for students, model schools, television programs, television and other publicity for 'good' school programs, more state and national examinations, more teacher training, longer school days and years, smaller classes, more computers in schools, more Internet connections, elimination of social promotions, standards and examinations for secondary school graduation, praying in schools, charter schools, vouchers, state control of public schools, private management of public schools, systemic reform, destruction of public schools and replacing them by private schools.

None of these cures has made any large difference in student learning, nationally or internationally. In the United States, states and the federal government have spent very large amounts of money on some of these approaches. Political philosophies rather than educational visions motivate some of these ideas. The reader will note that these ideas usually assume the information transfer paradigm for learning. Often they lead to additional costs to sustain them, funding usually not available.

13.5.2 United States reaction to Sputnik

The USSR Sputnik was a traumatic event for education in this country, already mentioned, leading to the belief that our educational systems, particularly in science and mathematics, needed improvement. We funded in response many large and well-supported curriculum efforts, mostly in the schools. One of us, Alfred Bork, worked on the Harvard Project Physics course for secondary schools, one of these efforts.

Just when we were getting good at developing such material (this was before computers were widely used in learning) one course destroyed the entire curriculum movement, Man, a Course of Study, a fifth grade sociology course. The proceedings of the extensive congressional investigation into this course are still interesting reading. The problems were entirely political, having nothing to do with students and learning. We have developed few large courses in the United States since then. The federal government never has recovered.

Even the new courses completed soon became minor players in the market. Often they were impractical for the typical learning situation. The material developed for elementary science was particularly interesting, with three major projects at the University of California, Berkeley, the Massachusetts Institute of Technology, and the American Association for the Advancement of Science. This material depended on discovery modules, not the memorization that dominated and still dominates the field. Students worked in groups of four or so. But these elementary science products were not widely used. Very few teachers, even with training, could manage such groups in typical school settings, although large sums of money went to teacher training. If we develop tutorial computer-based scientific literacy materials, these units will be valuable examples. Some examples already exist.

13.5.3 Solutions for Schools

The 1997 *Report to the President on the Use of Technology to Strengthen k-12 Education in the United States* is an important but neglected document. It devotes a major section to software content and pedagogy. It suggests a constructivist approach to learning software. Little such software exists for widespread use. The report says that:

> There is widespread agreement that one of the principle factors now limiting the extensive and effective use of technology within American schools is the relative dearth of high quality computer software

However, several years later we have little such software. The one response to this program known to us by the United States government emphasizes research based on present software, not new software.

13.5.4 Commercial competition with schools

With universities, as we will see, companies are already beginning to compete with them. But this happens less in the school market, perhaps because of the complexity of that market. Schools play many roles other than learning, often not well because of the mixture. One example is baby sitting, useful in a society where increasingly often both parents work.

An exception to major commercial investment in children's learning is the Edison Project, from Christopher Whittle. The original idea was to introduce a large number of private schools, heavily based on technology. We do not know the nature of the new learning material intended.

He hired well know people, including Benno Schmidt, President of Yale University, but apparently could not raise the funds for developing the technology based learning units necessary for his own schools. Then the plan changed, with more emphasis on existing schools, even public schools, and less on the new technology based curriculum material. The Edison approach now stresses efficient professional management of the schools. Recently Whittle has changed the name to Edison Schools and received sizable funding. There seems to be little current production of learning units, so the tutorial model is not the driving vision.

It is likely that we will soon see more private competition. Initially highly interactive tutorial units might be for use in traditional school settings and for home use. As units that are more effective are developed, we will begin to see more commercial competition. We mention one existing example soon.

13.5.5 Advanced placement

Problems that may be unique to the United States are the courses that prepare high school students for the advanced placement exams, possibly leading to university credit. Many different subject areas have such exams. High schools in wealthy areas offer many such courses. Poorer schools often cannot afford these courses for the limited number of students involved. This has already led to legal action by parents in these schools.

At least one organization, Apex Learning, is offering distance learning advanced placement courses. We do not know if these courses are tutorial. The URL for the Web site is http://www.apexlearning.com. For spring 2001, they list courses in mathematics, physics, chemistry, English, history, microeconomics, and politics. Others also exist.

13.5.6 Schools in poor areas

So far, this discussion about schools and the tutorial learning problems has referred mostly to the United States and other developed countries. Schools are an even more drastic problem in poorer areas, as already mentioned; often they are minimal, or do not exist. Western attempts to help have been ineffectual, based on trying to duplicate our poor efforts in parts of the world that will never be able to afford it at a sufficient level. Tutorial distance learning, in the students' native languages, using inexpensive learning appliances, could be a very important direction for improving learning in developing areas.

One intriguing possibility is that development of learning material for this market in wealthy areas could support the conversion to other languages and the distribution in the poorer areas. This is possible because these learning situations differ little from country to country.

> ..official primary school curricula are remarkably similar worldwide in terms of subjects taught and amount of time allocated to each
>
> Marlaine Lochhead and Henry Levin
> Effective Schools in Developing Countries
> Falmer, London, UK, 1993.

This would be a tremendous gift to the world! As in other areas, we need more experimental efforts with careful evaluation. Many languages would be necessary in this strategy.

13.6 Literacy

As indicated, early childhood education is critical. This continues past the preschool years. Often problems are common.

The results of the beginning school activities are various forms of literacy, needed for all later learning. Reading literacy gets the greatest attention, because, as stressed in the fictional accounts in Chapter 4, it is a key today to further study. However, mathematical literacy, writing literacy, scientific literacy, and visual literacy, among others, are also vital for the new century.

Reading and writing seem more specialized to the individual languages involved. We have more to learn about promoting visual literacy.

13.6.1 Science and mathematics

Mathematical literacy and scientific literacy present us with particularly interesting examples for initial development. The problems are international, and seem tractable to general worldwide solutions, allowing for cultural differences. However, little progress has been made.

> There are no valid reasons – intellectual, social, or economic – why the United States cannot transform its schools to make science literacy possible for all students. What is required is national commitment,

determination, and a willingness to work together toward common goals.

Science for All Americans
1989
Project 2061
American Association for the Advancement of Science

However, this has not happened, in the United States or elsewhere. Perhaps the problem is the emphasis on schools. Project 2061's recent studies have shown the inadequacy of science and mathematics textbooks in middle school. Yet these are critical to most courses offered. An interactive tutorial approach would be very desirable.

So we might begin with mathematics and science to gain the experience needed for the other literacies. The world is increasingly dependent on science, but even in developed areas, many people are scientifically illiterate.

13.6.2 Writing

Writing literacy is also interesting, seldom taught in schools, because of the difficulties of providing individual attention to each student in a class. Yet it is very important for modern communication.

There is little competition here. A few examples that are very labor intensive have been very successful, These could be further developed with tutorial material beginning with young children and continuing for many years. Voice input would be very valuable here, allowing a new approach to writing.

13.7 Informal learning environments

The home market may be important in moving us toward distance learning. Many parents are unhappy with their children's schools. They try to supplement school learning with home learning. There is already a market for computers in this activity. Preschool learning is one such large home market for highly effective carefully evaluated learning units.

Using computers in a variety of environments, such as public libraries, shopping centers, museums, community buildings, and other widely used locations shows strong promise. Alternatively, the learner could be anywhere, perhaps sitting in a field of daffodils or at the seashore. Clearly motivation is very important in such environments.

For shopping centers and similar locations, learning kiosks would be useful.

13.8 Apprentice learning

One variety of such learning, very old, is apprentice learning. A person learning how to make violins and cellos works with a master in this area, perhaps for many years. Many other examples of apprentice learning exist. Often on the job training is of this type. It is learning by doing. As in our precious discussions of simulations, the computer watches over the shoulder of the student.

This is a particularly appealing market, because there are few personal stakes that would lead people to fight a new approach. The reader should contrast this with the situation in schools and universities, where powerful conservative forces might resist any change. This approach would be a natural way to approach tutorial distance learning.

13.9 University education

University education will eventually be a major market for tutorial computer based learning. However as suggested many forces will resist change in universities. We begin this discussion with some major needs that suggest good beginning places.

13.9.1 Background courses.

One important area to consider for the United States is the large remedial courses, such as those in English and mathematics, with large numbers of students in almost all universities. The word 'remedial' is unpopular, but clear. This is material that 'should' have been learned in schools, but was not. Or the time interval since school was too great. On the other hand, perhaps the material has not been used and so has been forgotten. Similar problems can be found elsewhere.

The cost in a typical university for running these courses is considerable and the desire for teaching school subjects among research-oriented faculty is almost zero. To give some idea of the size of the market, we should survey these courses at many campuses, with particular attention to the costs for development and delivery.

Often students are in these courses because of an examination, such as an entrance exam for a calculus course. Many students cannot pass almost simplistic entrance exams, primarily high school mathematics, and are placed in remedial courses. Universities have oscillated in offering remedial courses, but the need for students to learn the type of material within them is indisputable.

Offering these courses for university students at community colleges is often unsuccessful. Many students who take these courses do not graduate, so nationally these courses have a poor reputation. For many students existing courses do not provide an adequate learning environment.

Student needs vary greatly from student to student. Often they need only a small portion of the material, but must take the whole boring course. An interactive tutorial course, reacting individually to each student, could be much more efficient in student time and more effective in student learning, as compared with the traditional courses in the area.

We once designed such a program in a joint project in remedial mathematics between the University of California and the California State University systems, based on a sizable empirical base from exams in five introductory subjects. We could not find funds for implementation.

These courses would be excellent for the trial experiment or later development. These badly needed courses with little market resistance would also furnish us with excellent marketing experience.

13.9.2 Large courses

Another similar educational problem and market exists for the large courses in universities. These courses are remarkably similar from university to university, because only a few textbooks dominate the market. They are also similar in other countries. In United States universities, about twenty-five such courses account for about half of the student credit hours at the universities.

These courses are almost inevitably poor, from the student viewpoint. They are given today as lecture-textbook courses, with a fixed pace and with no adjustment possible for individual students. Often in major universities, they are taught by lectures hired particularly for such courses, not by tenure track faculty. As with remedial courses, many research faculty do not want to teach these courses.

The amount of attention that individual students receive, often from teaching assistants, is not sufficient for many students in these courses. Far more students fail these courses or do poorly in them than should be the case. Oddly, many instructors and universities are proud of this weeding effect. They feel that it is their duty to fail students, an unfortunate attitude.

A tutorial technology-based course, highly interactive and adapting to the individual student needs, could establish important markets for any of the large courses. Because of the high percentage of students in these courses, these courses would challenge the economic basis of the university. Our experience in developing the introductory physics course at Irvine furnishes useful information for such development.

The computer could play an important role in any course normally taught with more that 100 students. The type of activities in these interactive courses could be far superior to that in the current courses of this size. Students could receive far more individualized attention from a tutorial program.

In developmental efforts, in the immediate future, we should probably avoid small courses or courses taught primarily with non-lecture approaches. However, most university courses are lecture based.

13.9.3 The future of universities

Not everyone believes that universities will survive.

> Thirty years from now the big university campuses will be relics. Universities won't survive. It's as large a change as when we first got the printed book. Do you realize that the cost of higher education has risen as fast as the cost of health care? ... Such totally uncontrollable expenditures, without any visible improvement in either the content or the quality of education, means that the system is rapidly becoming untenable. Higher education is in deep crisis... The college won't survive as a residential institution. "

Peter Drucker
Forbes 10 March 1997

Note that this is an economic prediction. Will the universities survive the onslaught of private distance learning companies?

If we act now with realism and foresight,
if we show courage,
if think globally and
allocate our resources accordingly,
we can give out children a
more peaceful and equitable world.
One where suffering will be reduced.
Where children everywhere
will have a sense of hope.
This is not just a dream.
It is our responsibility.

James D. Wolfenson
President, The World Bank
http://www.worldbank.org/ourdream/

Chapter 14

STARTING NEW DISTANCE LEARNING INSTITUTIONS

As we have often commented, no full scale tutorial computer-based distance learning programs exist. So we need much more experience and data to know that such a system will really work. Small-scale efforts indicate that this approach is very promising. If the research is favorable, we can proceed with this approach.

Several steps are necessary to implement a system based on tutorial computer based learning. Here is a tentative list; we consider many of these factors further in this chapter.

1. We need a planning group, or several planning groups, to arrange the following steps. A critical aspect will be to arrange the financial support for this development and for the full development that may follow. As suggested, some of this support might come from sales in wealthy areas.

2. We need addition research

- Voice input, particularly for very young children
- Delivering highly interactive learning through the Internet and satellites.
- Better research on tutorial learning, based on extensive major experiments to verify or refute the belief that computer-based tutorial learning is very important for the future.
- We lack adequate theories of learning; we may not even be close to effective predictive theories.

The tutorial computer-based learning materials, with their built in capabilities to store information frequently about the student, will assist with this research leading to vastly more knowledge about student learning. We have much to learn about learning.

3. We need many sequences in many areas and age groups.

   - Professional evaluations should test these sequences with large numbers of students in many environments.
   - This work should be separate from development, done by different people.
   - The computers can store much of the experimental data.

4. If the experimental efforts are successful, we should embark on full-scale development of new units. This will be a great endeavor for humanity, leading to new possibilities for the world.

Many development groups should participate. Different learning philosophies might be involved, reflecting our current weak knowledge of how people learn.

In a given area we should develop several sequences, developed by different groups to allow choice and to compare different learning approaches.

This activity needs continuous evaluation.

Many languages and cultures should be part of this major effort. International groups, reflecting different insights into learning should design and evaluate the learning segments.

Some of the organizations involved may be profit seeking.

We will need a continuing management system, for maintenance and for the development of new learning segments.

A professional evaluation process should be part of this activity.

4. Current schools, universities, and other learning institutions of today may not be adequate to lead us into the future

5. New organizations need to be created.

   - These include organizational structures for management, reporting, distribution, continuing maintenance and revision of the new learning material, including the development of new segments.

- We need more experience with tutorial computer-based material to guide the development of these entities. They may not resemble traditional educational organizations, but we should use the lessons of the past.

6. The system should be dynamic, allowing for new content and new learning strategies as our knowledge of how people learn increases.

In the next sections, we examine some of the details of these steps.

14.1 The experiment

We should begin with an experiment to show the effectiveness of the new approach proposed. The first question concerning the suggested major experiment on tutorial learning with computers is how to begin. Several approaches are possible, depending on the funding available. We could begin with a full experiment, or we could begin development directly.

14.1.1 Begin with the full experiment

In this desirable approach, we start with a most important experiment, looking to obtain extensive additional information before beginning large-scale development. Approximately twenty course-length learning segments in many areas and in several languages and cultures appear to provide an adequate empirical investigation.

Such variety would enable us to evaluate in many geographical locations and to gain empirical information about how these tutorial approaches compare with the existing material in the same area. We could experiment with different development approaches and explore marketing possibilities.

An experiment of this kind requires funding of about one hundred million dollars, over about five years. It could involve several stages, if only part of the money is initially available. Chapter 13 discusses the question of beginning courses.

14.1.2 The slow experiment

The second possibility is to begin immediately with individual modules, learning as we go. The development of each would depend on available funding. Here the initial necessary funding will be less, but less will be learned in this reduced experiment. The long-range profits might be delayed, because not much material would be quickly available.

Hence, we prefer the first alternative. However, finances might suggest the second. The next section discusses the proposed experiment.

14.1.3 Experimental details.

We need two separate groups, one to conduct the experiment and one to evaluate the experiment. We need first to develop the segments for the evaluation.

Only an outline of the steps is included. The developmental details are like those discussed for tutorial modules.

- Establish a planning group for the experiment.
- Secure adequate funding.
- Decide whether the choice of subject areas is by the planning group or through a solicitation for course proposals.
- Choose the subject areas.
- Develop the segments, using a variety of development procedures,
- Carefully gather information about costs and activities of each procedure.
- No single group should develop more than two modules, so a variety of learning styles should be represented.
- Begin at once to plan for full-scale development and evaluation.

14.1.4 Evaluation of the experiment.

To determine whether this experimental project is a success or not and to decide if we should start extensive development, studies need to be made in addition to the development of the material. We require several types of evaluation.

All the information gathered during the experiment, including the raw data, should be publicly available, perhaps on a Web site. This rich collection of data would allow others to make studies based on this information. We would expect extensive research outside of the experiment, conducted by many other groups.

Formative evaluations are part of the course developments themselves as in all well-done curriculum developments, as already described in our discussion of development in Chapter 11. The evaluation of the experiment discussed here is a separate activity.

14.1.4.1 Independent Evaluation

We should conduct, independent of the development of these materials, summative evaluations that compare these new tutorial learning units with those taught in traditional ways. We want strong evidence that students do learn more, learn faster, and retain information longer, as compared to the traditional learning methods. In addition, students should enjoy the new learning directions as compared to traditional learning. We are not looking for small statistical differences in learning; we hope that the new units wilt be *much more* effective than the old courses.

We also need to determine whether the learning units using technology-based tutors are economically efficient, costing less than lecture and other existing courses, yet still maintaining or improving quality of learning. This data will be essential for continuing development and widespread marketing. Large numbers of students need to be involved, both using the new and old approaches, and we require information about what works for what kind of student.

Skilled professional evaluators, independent of the other aspects of the project, should make these studies, so that they will be reputable from a scholarly point of view. We do not want the developers themselves, or their friends, simply announcing that everything worked wonderfully, so often found in education.

14.1.4.2 Evaluating marketing

Another aspect of the experimental studies in this first phase will be to study the questions of marketing. We hope that given a variety of levels and student audiences in the experimental units we can compare alternative marketing tactics and see which tactics are likely to be the most successful. There is little experience marketing full-scale, technology-based learning, so there is much to be learned.

Marketing will be assisted by the contrast of these materials with those using information transfer.

> What's wrong with the traditional educational model? According to [Jerry] Wind and [David] Reibstein [economics professors at Wharton] that model delivers standardized content, in a discrete time

and place, usually in a passive setting. In other words, a professor in a lecture hall imparts knowledge to a large number of students, who may be briefly engaged in discussions but are mostly passive. The lecturer also uses a one-size-fits-all approach, since the content of the lecture remains the same of each student, regardless of his or her individual needs. This model has worked well for centuries, however, in part because it is efficient for teachers. Its focus is on teaching rather than learning.

14.1.4.3 Decision to continue

The final task at this experimental stage will be deciding whether to continue to the larger project, full development. Early in the experimental phase, we need procedures for reaching this decision. Success in both improving learning and marketing will be important.

We will probably not be able to experiment with the possibility of full-scale distance learning at this stage, because too few modules of this type will be available. However, some work in this direction will be possible. We can verify that the new units will be fully usable in a distance learning environment.

As indicated, it may not be practical to conduct the full experiment described, perhaps because of the expense of such an experiment. In this case, an alternate approach, the slow experiment, is possible. We begin developing individual segments, one or several, as funds permit.

All the previous comments still apply in this case. We would eventually accumulate the knowledge that we would gain from the experiment, but at a slower rate.

14.2 FULL DEVELOPMENT

If we begin full development, a sizable activity, because the experiment was successful, we should already have a good basis of experience from the earlier experiment, both for the effectiveness of the courses and their appeal to the market.

The strategies for full development of each tutorial segment could be similar to those already described in Chapter 11. Some of the material will use international development, as in the experimental work. Learning insights from many countries will improve learning effectiveness. For full development we would expand the developmental tools already available and perhaps develop other approaches

14.2.1 Curriculum development factories

The production of many new courses, as would be required, is an extensive process, so we need to rethink strategies. One possibility is the creation of curriculum development factories for this purpose. Many technology-wise companies realize the advantages of specialized software groups.

These groups would specialize in the development of these new tutorial modules. Several groups would allow different ideas, but the groups should cooperate. They might be specialized as to area. Each might develop new tools to assist the process.

14.2.2 Distance learning again

We have discussed distance learning in Chapters 3 and 4, noting the variety of types available. A goal for distance learning is that the learning could take place at any place and at any time.

Liberating students from space and time considerations is an important step in making learning more available to all potential students. An essential feature in the notion of 'open university' is that all students could learn any subject. Interactive tutorial learning units, well constructed, can be available anywhere at any time, and can adjust to the level of the student, often allowing marketing at multiple levels. Here we stress tutorial computer-based learning as the model for the future.

Credit may or may not be involved. Students would have greater access to learning; we would gain many remote students.

The possibilities for creating new distance learning institutions based on computer technology are very interesting. The full use of highly interactive, multimedia tutorial learning modules is a new direction for distance learning institutions, very promising for the future.

14.2.3 Delivery

Several possibilities exist for the delivery of these new courses for distance learning, depending on the development of the technology in the next few years. The most likely possibility now is through CD-ROM or anetwork-based delivery procedures.. A combination of these (and possible new) technologies is likely for the future, as networks improve in speed and availability. As mentioned, satellite systems seem particularly attractive for future delivery, allowing us to reach everyone.

14.2.4 Areas of development

We have suggested in Chapter 13 some areas for developing tutorial modules using modern multimedia computers. All the lecture courses would be obvious possibilities for the early stages of full development.

Three factors are important in making choices, as stressed before.

- Improving student learning. Good courses of this type, stressing individualized student learning, could be of major help to all students.
- Opportunities for marketing. Good marketing possibilities for technology-based courses at the university level.
- Distance learning. We wish to look towards the formation of major distance learning institutions, based on highly interactive, multimedia learning units.

We are particularly concerned with materials that might well realize a profit in a few years, in spite of the sizeable development costs that would be involved. There would be a major market for such courses.

14.2.5 People

The developments suggested in this book would have major implications for teachers and faculty. They would also affect the structure of our schools and universities. This section reviews faculty implications.

14.2.5.1 Numbers

The new technology-based courses suggested in this document would require less human time than the existing lecture-textbook courses. The interactive courses already mentioned at Stanford University, in logic and set theory, give us information in this direction.

Patrick Suppes is nominally the instructor on these courses. However, he spends none of this time 'teaching' the courses. There are no lectures. A few teaching assistants work directly with the students, but most of the individualized student help comes directly from the computer.

The courses in the Open University and other distance learning institutes worldwide give us further information in this direction. The courses developed will be highly individualized.

14.2.6 ALL COURSES?

In some areas, it might be difficult to develop tutorial segments that could compete well with traditional learning. Only experience can establish this. However, we believe that anything currently taught in a large lecture environment can be taught in a much superior fashion with tutorial computer material.

On the other hand, we would expect in our vision that the courses given directly by professors in universities, if they survive, would be the smaller courses. These would have no more than approximately ten students, taught on a discussion basis or by other methods that go far beyond a lecture-textbook strategy. These courses would not be easy to replace with technology-based courses, given our present experience.

14.3 Marketing of Tutorial units

There will be a heavy demand for tutorial learning segments if careful research has shown that the new approach is far better for student learning than that currently available. We strongly suggest such research in this proposal, It is important both in establish the quality of the developed material and for success in marketing.

Furthermore, there would be a considerable market for informal use, outside institutions, for segments or entire courses. Many of these might also be sold to schools in this country and elsewhere in the world. Versions in other languages would be developed, as appropriate.

14.3.1 Gaining marketing experience

At present, as we emphasize there is little experience in marketing the types of courses considered. Hence, we should consider carefully the marketing issues, working with experts in marketing to the school, university, and other markets.

Considerations of this type should start right at the beginning of the projects, before there are any materials available to market. It should not be just an afterthought.

14.3.2 Developed and developing countries

We have emphasized that it is our hope that marketing in developed areas will lead to large profits, and that part of these profits can cover all or most of the cost for supplying learning to poor areas.

14.4 FUNDING

A major problem is to obtain funding, first for the experiment and then for the following extensive development.

To develop high quality interactive multimedia curriculum material of any type demands significant funding, as discussed in Chapter 12. This is up-front money, mostly. However, the eventual savings and profit can be great.

For the experiment, funding might come directly from governments or from organizations such as the United States National Science Foundation or other federal or state sources. However, we cannot count on this. Several countries might be involved. Commercial funding might be possible for full development, as eventually this will be a profitable area.

14.4.1 Government funding

We might persuade federal agencies or state legislatures to fund material of this kind, since it would probably save money in the end and even be a profitable venture. They might fund the experiment. If the experimental results are successful, further funding might come from other sources.

Governments concerned about learning of their citizens, in their schools and universities and elsewhere, could be funding curriculum development of tutorial learning materials. Funding such development is now rare in the United States, but more common elsewhere with the support of high quality large-scale distance learning institutions such as the United Kingdom Open University and the other mega-universities.

14.4.2 Cooperating institutions

Another possibility is that several schools, school districts, universities, or training centers might be interested in working together, each providing some of the funding. EDUCAUSE or similar organizations might facilitate this for universities.

14.4.3 Foundations

Private foundations interested in major improvements in higher education would be another possible source. The Alfred P. Sloan Foundation has supported asynchronous learning in universities, for example

14.4.4 Commercial funding

A likely direction to pursue for funding full development, we think, is organizations seeing this as a major source of profits. These might be new venture capital efforts or existing companies. We have many technology-rich companies interested in new areas to explore. The educational market is large, but so far technology is a minor component.

A major development would furnish an exciting new area for companies to explore, a market that no one is involved with to the extent suggested and one that has significant promise. Venture capital companies are beginning to realize the potentials for very large profits in computer-based learning. Among those involved in funding groups in this direction are Michael Milken and Paul Allen.

> Investors are pouring millions, soon to be billions, into the online education market. . . . Thomas Weisel Partners . . . estimate a $10 billion virtual higher education market by 2003 and an $11 billion corporate-learning market . . .
>
> Forbes
> The Virtual Classroom vs. the Real One
> http://www.Forbes.com/bow/2000/0911/bestofweb_print.html

These figures assume current information transfer online material. The potential should be much greater for tutorial units.

14.4.4.1 UNEXT

One promising profit-seeking group recently formed is UNEXT. It intends, we believe, to work at the university level.

It has a realistic idea of the funding and effort required for good development of learning units. UNEXT literature (http://www.unext.com/) emphasizes interactive learning, but we have not seen any of their material. Therefore, we do not know if their meaning for interaction is like that in this book.

Their most widely publicized effort has been in the business area. They have formed an allegiance with a number of prominent universities, including Stanford University, University of Chicago, Carnegie Mellon University, Columbia University, and the London School of Economics. We do not know just how these schools are involved. This is not the only direction for UNEXT.

14.4.4.2   Further commercial development

Profit companies in education are likely to be increasingly important in learning. This is a new and promising area. In the United States alone learning at all levels is a trillion-dollar industry. However, many will unfortunately work in the information transfer paradigm, as in almost all current online learning units.

Readers who have more experience with this might have more suggestions. All the areas discussed in this chapter are possibilities.

We should prepare a prospectus for venture capital groups and commercial companies, outlining the likely markets and the possibilities for working together in this direction. Perhaps this book would serve partially for that purpose.

14.4.5 Governments and Foundations

Governments concerned about learning of their citizens, in their schools and universities and elsewhere, could be funding development of tutorial learning materials for experiments in adult education. They funded other major experiments such as the United Kingdom Open University. Foundations could also be leaders in this effort. Both will be important in the initial experimental stages.

Governments concerned about learning of their citizens, in their schools and universities and elsewhere, could be funding curriculum development of tutorial learning materials. This is rare in the United States, but more common elsewhere with the support of high quality large-scale distance learning institutions such as the United Kingdom Open University and the other mega-universities.

14.4.6 International Organizations

The problems of learning, we have stressed, are worldwide. Many of the major problems of today, such as overcoming poverty, controlling population, and providing an adequate water supply for all, depend for their solution partially on effective learning for very large numbers in all areas. Learning itself is one of these major problems.

These problems are fundamentally international, not solvable in only one country. Organizations like UNESCO and the World Bank will be very important players in this time of transition.

They are already highly involved in learning, but often they are trying to promote the information transfer paradigm, modeling what they do after currently dominant approaches. We need to persuade them to try new directions that may be much more effective. They will be particularly important at the school level, because of the lack of adequate schools in developing areas.

We can see some progress in recent events. The World Bank has established sixteen distance learning centers in developing countries. It plans twenty more such centers.

Just as this book is going to press, Sir John Daniel, Vice Chancellor of the Open University, has been appointed the Assistant Director-General for Education for UNESCO, by the director-General Koichiro Matsuura. He will move to Paris in spring 2001. His experience with distance learning is likely to lead to major new efforts in UNESCO.

14.5 CONCLUSIONS

We summarize briefly in this final section some relevant information.

This book proposes the development of many highly interactive tutorial learning segments, replacing current approaches. The new paradigm for learning would be tutorial learning, replacing information transfer.

A full experiment would be this first step. This might further later large-scale development. Approximately twenty full segments would be developed for the experiment, and extensively tested against the existing lecture-textbook courses. About one hundred million dollars would be required for the full experiment. If the experiment is not successful, not leading to greatly improved student learning, work will stop. The alternative to the experiment would be to begin with individual courses, as funds are available.

If the experiment is successful, we can proceed with full scale development of new learning materials. The material would be full sequences in each subject area, usually multiple sequences, making extensive use of modern informatics technology. Media would appear as appropriate.   Distance learning would be the usual delivery method.

The new learning modules would be highly interactive, providing individualized help, matching the needs of each student, and keeping students interested.   They would resemble a learning situation involving a student and a skilled human tutor.

The courses could replace existing courses, in distance learning environments and in informal learning environments such as homes. Thus, there are many markets to pursue. For improving learning, and for marketing, careful evaluation is important.

> And indeed there will be time
> To wonder, Do I dare? And
> Do I dare?
> Time to turn back and descend the stair
> With a bald spot in the middle of my hair.
>   Do I dare
>   Disturb the Universe?

>   T.S.Eliot
>   The love Song of J. Alfred Prufrock

WE MUST DARE!

BIBLIOGRAPHY

1. Abruscato, Joe & Hazzard, Jack (1991). *The Whole Cosmos Catalog of Science Activities.* Glenview, IL: Good Year Books.
2. Ackoff, Russell Lincoln (1978). *The art of problem solving.* New York: Wiley.
3. Adams, James L. (1974). *Conceptual Blockbusting; A Guide to Better Ideas.* New York: W. H. Freeman.
4. Adams, James L. (1986). *The Care & Feeding of Ideas: A Guide to Encouraging Creativity.* Reading, MA: Addison-Wesley Pub. Co.
5. Akyalcin, Jeff (1997). *Constructivism - An Epistemological Journey from Piaget to Papert.* Melbourne, Australia: Kilvington Girls Grammar School.
6. Amabile, T.M. (1983). *The social psychology of creativity.* New York: Springer-Verlag.
7. American Association for the Advancement of Science. (1993). *Benchmarks for Scientific Literacy - Project 2061.* New York: Oxford University Press.
8. American Association for the Advancement of Science (1995). *Project 2061: Science Literacy for a Changing Future.* Washington, DC: American Association for the Advancement of Science, available at: http://www.project2061.org//
9. American Association for the Advancement of Science (1990). *The Liberal Art of Science: An Agenda for Action.* Washington, DC: American Association for the Advancement of Science.
10. Andrews, J.G. & McLone, R.R. (Eds.) (1976). *Mathematical modelling.* London, UK: Butterworths.
11. Anning, Angela (1997). *Early Childhood Education: A New Era?.* Staffordshire, UK: Trentham Books.
12. Apple, Michael & Beane, James (1995). *Democratic Schools.* Alexandria, VA: Association for Supervision and Curriculum Development.
13. Argyris, Chris (1982). *Reasoning, Learning, and Action: Individual and Organization.* San Francisco, CA: Jossey-Bass.
14. Arons, Arnold (1983). Achieving Wider Scientific Literacy. *Daedelus, 112(1).* pp. 91-122.
15. Arons, Arnold (1984). Computer-based Instructional Dialogs in Science Courses. *Science, 224(4653), June.*
16. Arons, Arnold (1977). *The Various Languages: An inquiry Approach to the Physical Sciences.* New York: Oxford University Press.
17. Arons, Arnold & Bork, Alfred (1964). *Science and Ideas: Selected Readings.* Englewood Cliffs, NJ: Prentice-Hall.
18. Ascher, Carol, Fruchter, Norm & Berne, Robert (1966). *Hard Lessons-Public Schools and Privatization.* New York: The Twentieth Century Fund Press.

19. Atkinson, Richard (1974). *Adaptive Instructional Systems, Technical Report 240.* Stanford University, CA: Institute for Mathematical Studies in the Social Sciences.
20. Atkinson, Richard C. (1965). *Scientific Psychology.* NY: Basis Books.
21. Atkinson, Richard C. (1976). Adaptive Instructional Systems: Some Attempts to Optimize the Learning Process. In D. Klahr (Ed.), *Cognition and Instruction* (pp. 81-108). Hillsdale, NJ: Lawrence Erlbaum Associates Publishers.
22. Atwell, Nancie (Ed.) (1990). *Coming to Know - Writing to learn in the intermediate grades.* Portsmouth, NH: Heinmann.
23. Banerji, Ranan B. (1969). *Theory of Problem Solving; An Approach to Artificial Intelligence.* New York: Elsevier.
24. Barker, Wayne (1968). *Brain Storms; A Study of Human Spontaneity.* NY: Grove Press.
25. Bates, A.W. (Ed.) (1990). Media and Technology in European Distance Education. *Proceedings of the EADTU Workshop on Media, Methods and Technology.* Heerlen, The Netherlands: EADTU.
26. Bean, John (1996). *Engaging Ideas - The professor's Guide to Integrating Writing, Critical Thinking and Active Learning in the Classroom.* San Francisco,CA: Jossey-Bass.
27. Beaton, A., Martin,M., Mullis, I., Gonzalez, E., Smith, T. & Kelly, D. (1996). Science Achievement in the Middle School Years: *IEA's Third International Mathematics and Science Study (TIMSS).* Boston College, Boston, MA: Center for the Study of Testing, Evaluation and Educational Policy.
28. Becker, Henry (1987). *Addressing the Needs of Different Groups of Early Adolescents: Effects of Varying School and Classroom Organizational Practices on Students from Different Social Backgrounds and Abilities. Report 16.* Baltimore, MD: Center for Research on Elementary and Middle Schools, Johns Hopkins University.
29. Becker, Henry & Ravitz, Jason (1998). The Equity Threat of Promising Innovations: Pioneering Internet-Connected Schools. *Journal of Educational Computing Research, 19.*
30. Becker, Henry, Ravitz, Jason & Wong,VanTien (1999). *Teacher and Teacher-Directed Student Use of Computers and Software.* U. S. Department of Education. NSF Grant # REC-9600614: University of California and University of Minnesota. available at: http://www.crito.uci.edu/tlc/findings/computeruse/
31. Belanger, P. & Valdivielso S. (1997). *The Emergence of Learning Societies: Who Participates in Adult Learning?.* Oxford, NY: Pergamon and UNESCO Institute for Education.
32. Belenky, Mary Fl. et al. (1986). *Women's Ways of Knowing: The Development of Self, Voice, and Mind.* NY: Basic Books.
33. Bellamy, Carol (2000). A Vision of Basic Education in the New Century. *TechKnowLogica, May/June,* available at: http://www.techknowlogia.org/TKL_active_pages2/TableOfContents/t-right.asp?IssueNumber=5
34. Bennett, William Ralph (1976). *Scientific & Engineering Problem-Solving with the Computer.* Englewood Cliffs, NJ: Prentice-Hall.
35. Berger, Dale E. (1987). *Applications of Cognitive Psychology: Problem Solving, Education, and Computing.* HillsdaleNJ: Lawrence Erlbaum Associates Publishers.
36. Berkowitz, Leonard (Ed.) (1978). *Group Processes.* New York: Academic Press.
37. Berliner, David & Calfee, Robert (Eds.) (1996). *Handbook for Educational Psychology.* New York: Macmillan Library Reference.
38. Billstein, Rick, Libeskind, Shlomo & Lott, Johnny W. (1987). *A Problem Solving Approach to Mathematics for Elementary School Teacher (3rd ed.).* Menlo Park, CA: Benjamin/Cummings.
39. Birkenholz, R. (1999). *Effective Adult Education.* Danville, IL: Interstate.

40. Block, James (1974). *Schools, Society, and Mastery Learning*. Holt, NY: Rinehart and Winston.
41. Bloom, Benjamin (1981). *All Our Children Learning: A Primer for Parents, Teachers, and Other Educators*. New York: McGraw Hill.
42. Bloom, Benjamin (1973). *Every Kid Can: Learning for Mastery*. Washington, DC: College/University Press.
43. Bloom, Benjamin (1984). The 2 sigma problem: The Search for Methods of Group Instruction as effective as one-to-one tutoring . *Educational Researcher, 13(6)*, 4-16.
44. Bloom, Benjamin, Hasting, Thomas & Madaus, George (1971). *Handbook on Formative and Summative Evaluation of Student Learning*. New York: McGraw Hill.
45. Bloom, Benjamin, Madaus, George & Hastings, Thomas (1981). *Evaluation to Improve Learning*. New York: McGraw Hill.
46. Boehm, E.H. & Horton, F.W. (1991). Distance Learning Methodology and Information Resources Management. *International Journal of Information Resource Managemen, 2(1)*.
47. Bork, Alfred (1999). An Interview with Alfred Bork - The Future of Learning. *EDUCOM Review, July/August*, available at: http://www.ics.uci.edu/~bork/papers.html.
48. Bork, Alfred (1987). Computer Networks for Learning. *Technological Horizons in Education Journal, 14(9)*.
49. Bork, Alfred (1995). Distance Learning and Interaction: Toward a Virtual Learning Institution. *Journal of Science Education & Technology, 4(3)*, 227-244.
50. Bork, Alfred (1988). Ethical Issues Associated with the Use of Interactive Technology in Learning Environments. *Journal of Research on Computing in Education, 21(2)*.
51. Bork, Alfred (1996). Highly Interactive Multimedia Technology and Future Learning. *Journal of Computing in Higher Education, 8*, 1-26.
52. Bork, Alfred (1987). Interaction: lessons from computer-based learning. In Laurillard, D. (Ed.), *Interactive Media Working Methods and Practical Applications*. Chinceter, UK: Ellis Horwood Ltd.
53. Bork, Alfred (2000). Learning Technology. *EDUCAUSE Review. January/February*.
54. Bork, Alfred (1981). *Learning with Computers*. Bedford, Mass: Digital Press.
55. Bork, Alfred (1986). *Learning with Personal Computers*. New York:Harper & Row.
56. Bork, Alfred (2000). Learning with the World Wide Web. *The Internet and Higher Education*.
57. Bork, Alfred (1988). New Structures for Technology-Based Courses. *Education and Computing* , 109-117.
58. Bork, Alfred (1985). *Personal Computers for Education*. also available in Spanish and Japanese: Harper and Row.
59. Bork, Alfred (1987). Planning for the Future of Education. *Technology and Learning, 1(5), September/October*.
60. Bork, Alfred (1989). Production of Technology-Based Learning Material Tools vs. Authoring Systems. *Instruction Delivery Systems, 3(2), March/April*.
61. Bork, Alfred (1996). Rebuilding Universities with Highly Interactive Multimedia Curriculum. *International Journal of Engineering Education, 12*, available at: http://www.ics.uci.edu/~bork/papers.html
62. Bork, Alfred (1993). Schools for Tomorrow. *International Journal of Educational Research, 19(2)*.
63. Bork, Alfred (1993). Technology in education: An historical perspective. In Muffoletto, R. & Knupfer, N. N. (Eds.) *Computers in education: Social political for historical perspectives* (pp. 71-90). Cresskill, New Jersey: Hampton Press Inc.
64. Bork, Alfred (1997). The Future of Computers & Learning. *T. H. E. Journal, 24(11)*, June. 69.

65. Bork, Alfred (1987). The Potential for Interactive Technology. *Byte, February.* 201.
66. Bork, Alfred (2001). Tutorial Learning for the 21st Century. In press: *Journal of Science Education and Technology.*
67. Bork, Alfred (1995). Why Has the Computer Failed in Schools and Universities. *Journal of Science Education and Technology, 4(2).* 97-102.
68. Bork, Alfred & Ibrahim Bertrand, et. al. (1992). The Irvine-Geneva Course Development System. In Aiken, R. (Ed.) *Education and Society Information Processing 91, Vol 11.* BV, North Holland: Elsevier Science Pub.
69. Bork, Alfred & Weinstock, Harold (Eds) (1986). *Designing Computer Based Learning Material.* New York: Springer-Verlag.
70. Bork, Alfred, Britton, David & Gunnarsdottir, Sigrun (1994). Combining learning and assessment. In Beattie, K., McNaught, C. & Wills, S. (Eds.), *Interactive Multimedia in University education: Designing for change in Teaching and Learning* (A-59) (pp. 113-129). North-Holland: Elsevier Science B.V.
71. Bork, Alfred, Walker, David & Poly, Andre (1992). *Applications In Education and Informatics Worldwide.* Unesco: Jessica Kingsley.
72. Bourgeos, E. , Duke, C. , Guyot, J. L. & Merrill, B. (1999). *The Adult University.* Buckingham: The Society for Research into Higher Education and Open University Press.
73. Bransford, John & Stein,Barry S. (1984). *The Ideal Problem Solver: A Guide for Improving Thinking, Learning, and Creativity.* New York: W. H. Freeman.
74. Bransford, John, Brown, Ann & Cocking, Rodney (1999). *How People Learn: Brain, Mind, Experience and School.* Washington, DC: National Academy Press.
75. Brooks, Jacqueline & Brooks, Martin (1993). *In Search of Understanding: The Case for Constructivist Classrooms.* Alexander,VA: Association for Supervision and Curriculum Development.
76. Brown, E. L. (1993). *A Change Process of Developing Exemplary Middle Schools.* Universty of Southern California:dissertation.
77. Brown, J. C. (1970). *The Troika incident .* New York: Doubleday and Co. Inc.
78. Brown, Stephen I. & Walter,Marion I. (1983). *The Art of Problem Posing.* Philadelpha, PA: Franklin Institute Press.
79. Bruce, Bertram & Levin, James (1997). Educational Technology: Media for Inquiry, Communication, Construction, and Expression. *Journal of Educational Computing Research, 17.*
80. Bruner, Jerome (1990). *Acts of Meaning.* Cambridge, MA: Harvard University Press.
81. Bruner, Jerome (1996). *The Culture of Education.* Cambridge, MA:Harvard University Press.
82. Bruner, Jerome & Haste, Helen (1987). *Making Sense: The Child's Construction of the World.* NY: Methuen.
83. Buchanan, Elizabeth (2000). Emerging Ethical Issues in Distance Education. *The CPSR Newsletter, 18(2),* available at:
http://www.cpsr.org/publications/newsletters/issues/2000/Spring2000/buchanan.html
84. Burge, Elizabeth & Roberts, Judith (1993). *Classrooms with a Difference: A Practical Guide to the Use of Conferencing Technologies.* University of Toronto Press: Ontario Institute for Studies in Education.
85. Bybee, R. W. , Buchwald, C. E. , Crissman, S. , Heil, D. R. , Kuerbis, P. J. , Matsumoto, C. & McInerney, J. (1990). *Science & Technology Education for the Middle Years: Frameworks for Curriculum & Instruction.* Washington, DC: The National Center for Improving Science Education, The Network, Andover, MA and the Biological Sciences Curriculum Study.
86. Cahoon, B. (1998). *Adult Learning & the Internet.* San Francisco, CA: Jossey-Bass.

87. Calfee, Robert (1994). *Implications of Cognitive Psychology for Authentic Assessment.* Washington, DC: US Department of Education.
88. California State Department of Education (1998). *Challenges, Opportunities, Changes.* Final Report. Sacramento CA. Desember: California State Department of Education, Joint Board Task Force Non-Credit and Adult Education, available at: http://www.otan.dni.us/webfarm/jbtf/
89. California State Board of Education (1998). *Science Content Standards for Claifornia Public Schools. Kindergarten through grade twelveK-12 as recommend to the State of California.* California State Board of Education, available at: http://www.cde.ca.gov/board/pdf/science.pdf
90. Campbell, Patricia & Storo, Jennifer (1999). Reducing the Distance: Equity Issues in Distance Learning in Public Education. *Journal of Science Education and Technology, 3(4).*
91. Carnegie Council on Adolescent Development (1989). *Task Force on Education of Young Adolescents. Turning Points: Preparing American Youth for the Twenty-First Century.* New York: Carnegie Corporation.
92. Castro, Claudio de Mura, et al. (1998). Making Education a Catalyst for Progress . *The Contribution of the Inter-America Development Bank. April.*
93. Cazden, Coutney (1988). *Classroom Discourse - The Language of Teaching and Learning.* Portsmouth, NH: Heinemann.
94. Chambliss, Marilyn & Calfee, Robert (1998). *Textbooks for Learning: Nurturing Children's Minds.* Malden, MA: Blackwell.
95. Chance, Paul (1986). *Thinking in the Classroom: A Survey of Programs.* Columbia University, NY:Teachers College.
96. Ching, Cynthia Carter, KaFAI, Yasmin & Marshall, Sue (2000). Spaces for Change: Gender and Technology Access in Collaborative Software Design. *Journal of Science Education and Technology, 9.*
97. Clark, John (1990). *Patterns of Thinking.* Boston, MA: Allyn and Bacon.
98. Clarke, A. (1956). *The City & the Stars.* Brace and World: Harcourt.
99. Claxton, Charles & Murrell, Patricia (1987). Learning Styles: Implications for Improving Educational Practices. *ASHE-ERIC Higher Education Report, 4.*
100. Colby, Benjamin N. & Knaus, Rodger (1973). *Divination and Narrative Problem-Solving.* Irvine, CA: University of California.
101. Cole, Michael (1996). *Cultural Psychology: A Once and Future Disciple.* Cambridge, MA: Harvard University Press.
102. Collins, Cathy & Mangiere, John (1992). *Teaching Thinking: An Agenda for the Twenty-first Century.* Hillsdale, NJ: Lawrence Erlbaum Associates Publishers.
103. Collis, B., Knezek, G., Lai, K. W., Miyashita, K. T., Pelgrum, W., Plomp, T. , & Sakamoto, T (1996). *Children and computers in school.* Mahwah, NJ: Lawrence Erlbaum Associates Publishers.
104. Conyne, Robert K. (Ed.) (1985). *The Group Workers' Handbook: Varieties of Group Experience.* Springfield, IL: C.C. Thomas.
105. Costa, Arthur L. (1991). *Developing Minds: A Resource Book for Teaching Thinking.* Alexandria,VA: Association for Supervision and Curriculum Development.
106. Cronback, Lee & Gagne, Robert (Ed.) (1967). *How Can Instruction be Adapted to Individual Distances in Learning and Individual Differences.* NY: Merrill.
107. Daniel, Sir John (1996). *Mega-Universities and Knowledge Media.* London, UK: Kogan Page.
108. Daniel, Sir John (1997). Why Universities Need Technological Strategies. *Change, July/August.*

109. Daniels, Harry (Ed.) (1996). *An Introduction to Vygotsky.* London, UK: Rutledge.

110. Daniels, Harry (Ed.) (1993).*Charting the Agenda: Educational Activity after Vygotsky.* London, UK: Rutledge.

111. Darling-Hammond, Linda (1997). *The Right to Learn: A Blueprint for Creating Schools that Work.* San Francisco, CA: Jossey-Bass.

112. De Bono, Edward (1972). *Children Solve Problems.* London, UK: Penguin Press.

113. Dougherty, John W. (1997). *Four Philosophies that Shape the Middle School.* Bloomington, IN: Phi Delta Kappa Educational Foundation.

114. Drucker, Peter (1993). *Post-Capitalist Society.* New York: Harper-Collins.

115. Duckworth, Eleanor, Easley, Jack, Hawkins, David & Henriques, Androula (1990). *Science Education: A Minds-on Approach for the Elementary Years.* Hillsdale, NJ: Lawrence Erlbaum Associates Publishers.

116. Edwards, R. (1997). *Changing Places: Flexible Lifelong Learning and a Learning Society.* NYC, New York: Routledge.

117. Eichhorn, D. H. (1987). *The Middle School.* Columbus, OH: National Middle School Association.

118. Elliot, S. N. (1991). Authentic assessment: an introduction to a neobehavioral approach to classroom assessment. *School Psychology Quarterly, 6(4),* 273-278.

119. Evans, Jonathan (Ed.) (1983). *Thinking and Reasoning: Psychological Approaches.* London, UK: Routledge and Kegan Paul.

120. Farhad, S. (2000). Research in Distance Education: A Status Report. *International Review of Research in Open and Distance Learning, 1, June.*

121. Feldmann, S & Elliot, G. (1990). *At the Threshold: The Developing Adolescent.* Cambridge, MA: Harvard University Press.

122. Five, Cora Lee & Dionisio,Marie (1996). *Bridging the Gap: Integrating Curriculum in Upper Elementary and Middle Schools.* Portsmouth, NH: Heinemann.

123. Forman, George & Pufall, Peter (1988). *Contructivism in the Computer Age.* Hillsdale, NJ: Lawrence Erlbaum Associates Publishers.

124. Fox, William M. (1987). *Effective Group Problem Solving: How to Broaden Participation, Improve Decision Making and Increase Commitment to Action.* San Francisco, CA: Jossey-Bass.

125. Gabel, Dorothy (1994). *Handbook of Research on Science Teaching & Learning.* New York: Macmillan Library Reference.

126. Gage N. L. & Berliner, D. C. (1988). *Educational psychology(4th ed.)* Boston, MA: Houghton Mifflin Company.

127. Gardiner, A. (1987). *Discovering Mathematics: The Art of Investigation.* New York: Oxford Clarendon Press.

128. Gardner, Martin (1978). Aha! Insight. NY: W.H. Freeman & Co.

129. Gayol, Yolanda & Schied, Fred (1999). *Cultural Imperialism in the Virtual Classroom: Critical Pedagogy in Transnational Distance Education.* University Park, PA: Penn State University, available at: http://www.geocities.com/Athens/Olympus/9260/culture.html

130. Gentner, Dedre (1990). *Metaphor as Structure Mapping the Relational Shift, Technical Report 488.* University of Illinois at Urbana-Champaign: Center for the Study of Reading.

131. Gentner, Derde & Stevens,Albert (1983). *Mental Models.* Hillsdale, NJ: Lawrence Erlbaum Associates Publishers.

132. George, Paul, Lawrence, Gordon & Bushnell, Donna (1998). *Handbook for Middle School Teaching.* New York: Longman.

133. Gilhooly, K. J. (1982). *Thinking: Directed, Undirected, and Creative.* New York: Academic Press.

134. Glaser, Robert & Cooley, Willam (1978). *Research & Development & School Change: A Symposium of the Learning Research & Development Center.* Hillsdale, NJ: Lawrence Erlbaum Associates Publishers.

135. Goodyear, Peter (2000). *eLearning, Knowledge Work and Working Knowledge.* published on the IST2000 website in the eLearning Futures session http://istevent.cec.eu.int/en/details. asp?session=33&lang=eng. *November.* also available at http://domino.lancs.ac.uk/edres/csaltdocs.nsf

136. Graves, Donald (1991). *Build a Literate Classroom.* Portsmouth, NH: Heinemann.

137. Green, K. & Gilbert, S. (1995). Great expectations: content, communications, productivity, and the role of information technology in higher education. *Change Magazine, March-Apri,.* available at: http://www.virtualschool.edu/mon/Academia/GilbertSlideRulesAndProductivit

138. Gunderson, Keith (1985). Mentality and machines. MN: University of Minnesota Press.

139. Hare, Alexander Paul (1982). *Creativity in Small Groups.* Beverly Hills, CA: Sage Publications.

140. Harmin, Merrill (1994). *Inspiring Active Learning: A Handbook for Teachers.* Alexandria,VA: Association for Supervisors and Curriculum Development.

141. Harper, Barry & Hedberg, John (1997). *Creating Motivating Interactive Learning Environments: A Constructivist View.* New South Wales, Australia: University of Wollongong, available at: http://www.curtin.edu.au/conference/ascilite97/papers/Harper/Harper.html

142. Harry, Keith (1999). *Higher Education Through Open and Distance Learning.* London, UK: Routledge.

143. Hayes, John (1981). *The Complete Problem Solver.* Philadelpha, PA: Franklin Institute Press.

144. Heiman, Marcia & Slomianko, Joshua (1985). *Critical Thinking Skills.* Washington, DC: National Education Association.

145. Hemmerich, Hal, Lim, Wendy & Neel, Kanwal (1994). *Primetime! Strategies for Lifelong Learning in Mathematics and Science in the Middle and High School Grades.* Portsmouth, NH: Heinemann.

146. Hess, Jacqueline (1994). Learning to Ask the Right Questions. *EDUCOM Review, 29(3).*

147. Hiltz,S. R. (1994). *The Virtual Classroom.* New Jersey: Ablex Publishing Corporation.

148. Hodgson, Barbara (1993). *Key Terms and Issues in Open and Distance Learning.* London, UK: Kogan Page.

149. Holton, Gerald (2000). *Coupling Science and the National Interest.* Lecture at Harvard University, available at: http://www.ksg.harvard.edu/iip/lmb/holton.htm

150. Hoy, Wayne (1997). *The Road to Open and Healthy schools : a Handbook for Change Middle and Secondary School.* Thousand Oaks, CA: Corwin Press.

151. Hoy, Wayne & Sabo, Dennis (1998). *Quality Middle Schools.* Thousand Oaks, CA: Corwin Press.

152. Hunt, Earl B. , Marin, Janet & Stone, Philip J. (1966). *Experiments in Induction.* New York: Academic Press.

153. Hunt, Gilbert, Wiseman, Dennis & Bowden, Sandra (1998). *The Middle Level Teachers' Handbook: Becoming a Reflective Practitioner.* Springfield, IL: C.C. Thomas.

154. Hurd, Paul DeHart (1997). Inventing Science Education for the New Millennium. New York: Teachers College Press.

155. Illich, Ivan (1970). *Deschooling Society.* New York: Harper and Row.

156. Illich, Ivan (1973). *Tools for Conviviality.* New York: Harper and Row.

157. Immegart, Glenn L. & Boyd, William Lowe (1979). *Problem-finding in Educational Administration.* Lexington, MA: Lexington Books.

158. Johnson, E. (1998). *Linking the National Assessment of Educational Progress and the Third International Mathematics and Science Study: A Technical Report.* Washington, DC: United States Department of Education, National Center for Educational Statistics, available at: http://nces.ed.gov/pubs98/linking98/

159. Johnson, Lawrence, LaMontagne, M. J., Elgas, Peggy & Bauer, Ann (1998). *Early Choldhood Education - Blending Theory, Blending Practice.* Baltimore, MD: Paul H. Brooks Publishing.

160. Jones, Ann, Kirkup, Gil & Kirkwood, Adrian (1993). *Personal Computers for Distance Education.* New York, NY: St. Martin's Press.

161. Jones, Ann, Scanlon, E. & O'Shea, T. (Eds) (1987). *The Computer Revolution in Education: New Technologies for Distance Teaching.* New York, NY: St. Martin's Press.

162. Jones, Beau Fly, Palincsar, Annemarie Sullivan, Ogile, Donna Sederburg, Carr & Eileen Glynn (1987). *Strategic Teaching and Learning: Cognitive Instruction in the Content Areas.* Alexandria,VA:Association for Supervision and Curriculum Development.

163. Jones, Glenn (1991). *Make All America a School: Mind Extension University.* Englewood, CO: Jones 21st Century.

164. Kahney, Hank (1986). *Problem solving: A Cognitive Approach.* Milton Keynes, UK: Open University Press.

165. Karplus, Robert (1980). Teaching for Development of Reasoning. In Lawson, A. (Ed.), *The Psychology of Teaching and Creativity.* Columbus, OH: ERIC-SMECA.

166. Karplus, Robert & Their, Herbert (1967). *A New Look at Elementary School Science.* Chicago: Rand Mcnally.

167. Kaufmann, Geir (1979). *Visual Imagery and its Relation to Problem.* New York: Columbia University Press.

168. Keiser, K., Nelson, J., Norris, N. & Szyszkiewicz S. (1998). *NAEP 1996 Science Cross-State Data Compendium for the Grade 8 Assessment* - Findings from the National Assessment of Educational Progress for the State Science Assessment. Washington, DC: US Department of Education.

169. Keller, Fred (1968). Goodbye Teacher. *Journal of Applied Behaviorial Analysis. 1,* 79-89.

170. Keller, J. M. & Suzuki, K. (1988). Use of the ARCS motivation model in courseware design. In D. H. Jonassen (Ed.) *Instructional designs for microcomputer courseware.(* pp. 401-434). Hillsdale, NJ: Lawrence Erlbaum Associates Publishers.

171. Kellough, R. , Collette, A. , Chiappette, E. , Souviney, R. ,Trowbridge, L. & Bybee, R. (1996). *Integrating Mathematics and Science for Intermediate and Middle School Students.* Columbus, OH: Prentice Hall.

172. Kellough, Richard & Kellough, Noreen (1996). *Middle School Teaching : A Guide to Methods and Resources (2nd ed.).* Englewood Cliffs, NJ: Prentice Hall.

173. Kiesler, Sara & Turner, Charles (1977). *Fundamental Research and the Process of Education.* Washington, DC: Committee on Fundamental Research Relevant to Education, National Academy of Sciences.

174. Kirst, Michael (1987). *Evaluating State Educational Reforms: A Special Legislative Report.* National Conference of State Legislators.

175. Klahr, David (Ed.) (1976). *Cognition and Instruction.* Hillsdale, NJ: Lawrence Erlbaum Associates Publishers.

176. Kleinmuntz, Benjamin (1966). *Problem Solving: Research,Method and Theory.* NY: Wiley.

177. Korsgaard, O. (1997). *Adult Learning and the Challenges of the Twenty-first Century.* Odense: Association for World Education/Odense University Press.

178. Kozma, R. J. (1991). Learning with Media. *Review of Educational Research, 61(1),* 179-211.
179. Kozulin, Alex (1990). *Vygotsky's Psychology - A Biography of Ideas.* NY: Harvester Wheatsheaf.
180. Kuhn, Thomas (1998). *The Structure of Scientific Revolutions (Third edition).* Chicago: University of Chicago Press.
181. Lajoie, S. P. (1993). Computer environrnments as cognitive tools for enhancing learning. In Lajoie, S. P. & Derry, S. J. (Eds.) *Computers as cognitive tools* (pp. 261-288). Mahwah, NJ: Lawrence Erlbaum Associates Publishers.
182. Laurillard, Diana (1993). *Rethinking University Teaching.* London, UK: Routledge.
183. Leonard, G. B. (1968). *Education & Ecstasy.* New York: Delacorte Press.
184. Leonard, G. B. (1981). *Education & Ecstasy.* Berkeley, CA: North Atlantic Books.
185. Lerner, R. (1993). *Early Adolescence: Perspectives on Research Policy and Intervention.* Hillsdale, NJ: Lawrence Erlbaum Associates Publishers.
186. Lester; Frank K. (Ed.) (1982). *Mathematical problem solving : issues in research.* Philadelpha, PA: Franklin InstitutePress.
187. Levin, Henry & Lockheed, Marlaine (1993). *Effective Schools in Developing Countries.* London, UK: Falmer.
188. Lewis, A. C. (1991). *Gaining Ground: The Highs and Lows of Urban Middle School Reform 1989-1991.* New York: Edna McConnell Clark Foundation.
189. Lewis, J. L. & Kelley, P. J. (1987). *Science and Technology Education and Future Human Needs.* Oxford, NY: Pergamon Press.
190. Linn, Marcia & His, Sherry (2000). *Computers, Teachers, Peers - Science Learning Partners.* Mahwah, NJ: Lawrence Erlbaum Associates Publishers.
191. Lochhead, Jack (1987). *Thinking : the second international conference Perkins, D. N. (Ed.).* Hillsdale,NJ:Lawrence Erlbaum Associates Publishers.
192. Lockheed, Marlaine & Longford, Nicholas (1989). *A Multi-level Model of School Effectiveness in a Developing Country.* Washington, DC: Population and Human Resources Department, The World Bank.
193. Lockheed, Marlaine & Rodd, Alastair (1991). *World Bank Lending for Educational Research 1982-89.* Washington, DC: Population and Human Resources Department, The World Bank.
194. Lockheed, Marlaine & Verspoor, Adriaan (1991). *Improving Primary Education in Developing Countries.* Oxford, UK: Oxford University Press.
195. Lockheed, Marlaine, Hunter, B., Anderson, R., Beazley, R. & Esty, E. (1983). *Computer Literacy - Definition and Survey Items for Assessment in School.* Washington, DC: National Center for Education Statistics, United States Department of Education.
196. Longworth, Norman & Keith, Davcs W (1996). *Lifelong Learning - New vision, new implications; new roles for people, organizations, nations, and communities in the 21st century.* London, UK: Kogan Page.
197. Loree, M. Ray (1965). *Problem-solving techniques of children in grades four through nine.* University of Alabama.
198. Loucks-Horsley, S., Brooks, J. G., Carlson, M., Kuerbis, P., Marsh, D., Padilla, M., Pratt, H. & Smith,K. (1990). *Developing and Supporting Teachers for Science Education in the Middle Years.* Andover, Mass: The National Center for Improving Science Education.
199. Maier, Norman & Frederick, Raymond (1970). *Problem solving and creativity in individuals and groups.* Belmont, CA: Brooks/Cole Pub. Co.
200. Malone, Mark. (1987). *Physical Science Activities for Elementary and Middle School.* Washington, DC: National Academy Press.

201. Maryland, Laurel (1985). The Role of language in problem solving. In edited proceedings, Bruce, Robert Jernigan (Ed.) of the *symposium held at the Johns Hopkins University Applied Physics Laboratory (29-31 October 1984)*. Amsterdam, New York: Elsevier Science Pub. Co.

202. Mason, R. & Kaye, A. (1989). *Mindweave*. Oxford, NY: Pergamon.

203. Mayer, Richard E. (1983). *Thinking, Problem solving, Cognition.* New York: W. H. Freeman and Company .

204. Mayes, J. Terry (1995). Learning Technologies & Groundhog Day. In W. Strang, V. B. Simpson & D. Slater, *Hypermedia at work: Practice and theory in Higher Education.* Canterbury: University of Kent Press, available at: http://cad017.gcal.ac.uk/clti/staff/TMpapers.html

205. McClintock, Robbie (1999). *The Educators Manifesto - Renewing the Progressive Bond with Posterity through the Social Construction of Digital Learning Communities.* Columbia University, NY: Institute de Learning Technologies, Teachers College, available at: http://www.ilt.columbia.edu/Publications/manifesto/index.html

206. McEwin, C. Kenneth, Dickinson, Thomas S. & Jenkins, Doris M. (1996). *America's Middle Schools: Practices and Progress: A 25 Year Perspective.* Columbus, OH: National Middle School Association.

207. McGuire, Saundra (1989). *Elementary and Middle School Science Improvement Project.* Washington, DC: National Aeronautics and Space Administration.

208. McKay, Jack (1995). *Schools in the Middle: Developing a Middle-Level Orientation.* Thousand Oaks, CA: Corwin Press.

209. McKim, Robert H. (1980). *Thinking visually : a strategy manual for problem solving .* Belmont, CA: Lifetime Learning Publications.

210. McLean, L. D. (1990). Time to replace the classroom test with authentic measurement. *The Alberta journal of Educational Research, 36(1).* 78-84.

211. McNeil, Patricia W. (1996). *Preparing Youth for the Information Age: A Federal Role for the 21st Century.* Washington, DC: Institute for Educational Leadership.

212. Meyer, C. A. (1992). What's the difference between authentic and performance assessment. *Educational Leadership. 49(8).* 39-40.

213. Milken Family Foundation (1999). *Poll Reveals High School Students Saying "No" to Teaching Careers.* Milken Family Foundation. *36329,* available at: http://www.mff.org/newsroom/news.taf?page=72&type=archive

214. Minicucci, Catherine (1996). *Learning Science and English - How School Reform Advances Scientific Learning for Limited English Proficient Middle School Students.* Washington, DC: National Center for Research On Cultural Diversity and Second Language Learning.

215. Moll, Luis (Ed.) (1990). *Vygotsky and Education - Instructional Implications and Applications of Sociohistorical Psychology.* Cambridge, UK: Cambridge University Press.

216. Moore, Carl M. (1987). *Group techniques for idea building.* Newbury Park: Sage Publications.

217. Moursund, David G. (1986). Computers and problem solving, a workshop for educators. International Council for Computers in Education: Eugene, Ore.

218. Muth, K. Denise & Alvermann, Donna E. (1999). *Teaching and learning in the Middle Grades.* Boston, MA: Allyn and Bacon.

219. NAEP Science Consensus Project (1999). Science Framework for the 1996 and 2000 National Assessment of Educational Progress. Washington, DC: National Assessment Governing Board.

220. Napier, Rodney. W. (1985). *Group, theory and experience. (3rd ed).* Boston, MA: Houghton Mifflin Company.

221. National Center for Educational Statistics (1999). *NAEP 1998 Writing - Report Card for the Nation and the States.* September 1999: U. S. Department of Education.

222. National Council of Teachers of English (1996). Motivating Writing in Middle School. Urbana, IL: National Council of Teachers of English, available at: http://www.ncte.org/books/98/NCTE52872.html

223. National Research Council (1999). *Improving Student Learning - A Strategic Plan for Educational Research and its Utilization.* Washington, DC: National Academy Press.

224. National Research Council. (1997). *Improving Student Learning in Mathematics and Science - The Role of National Standards in State Policy.* Washington, DC: National Academy Press.

225. National Science Board. (1999). *Preparing our Children* - Math and Science Education in the National Interest. NSB 99-31. *March.*

226. National Science Foundation (1996). Women, Minorities and Persons with Disabilities in Science and Engineering. Artington, VA: NSF 00-327, available at: http://www.nsf.gov/sbe/srs/nsf00327/start.htm

227. National Science Resources Center (1988). *Resources for Teaching Middle School Science.* Washington, DC: National Academy Press.

228. National University Continuing Education Association (NUCEA) (1993). *The Electronic University: A Guide to Distance learning.* Princeton, NJ: Peterson's Guides .

229. Newell, Allen & Simon, Herbert A. (1972). *Human problem solving.* Englewood Cliffs, NJ: Prentice-Hall.

230. Nickerson, Raymond S. , Perkins, David N. & Smith, Edward E. (1985). *The teaching of thinking.* Hillsdale, NJ: Lawrence Erlbaum Associates Publishers.

231. NMSA (1979). *Middle School Research: Selected Studies 1977-1979.* Fairborn, OH: National Middle School Association.

232. Noam, Eli M. (1995). Electronics and the dim future of the university. *Science. 270, 13 Oktober. (*pp. 247-249), available at: http://www.uta.fi/FAST/JH/noam.html

233. Norman, Donald (1998). *The Invisible Computer.* Cambridge, Mass.: MIT Press.

234. Novak, Joseph & Gowin, D. Bob (1984). *Learning How to Learn.* New York: Cambridge University Press.

235. Oaks, J. (1990). *Multiplying Inequalities: The Effects of Race, Social Class, and Tracking on Opportunities to Learn Science and Mathematics.* Santa Monica, CA:Rand Corporation.

236. Oblinger, Diana (1998). Technology and Change: Impossible to Resist. *NCA Quarerly. 72.*

237. Olsen, Shirley A. (Ed.) (1982). *Group planning and problem-solving methods in engineering management.* New York: Wiley.

238. Olson, Carol Booth (1996). *Practical Ideas for Teaching Writing at the Elementary School and Middle School Levels.* Sacramento, CA: California Department of Education.

239. Olson, Carol Booth (1992). *Thinking/Writing: Fostering Critical Thinking Through Writing.* New York: Harper-Collins.

240. Orginazation for Economic Co-operation and Development (1986). *Information Technologies and Basic Learning.* Paris: Centre for Educational Research and Innovation.

241. Overholt, James L., Rincon, Jane B. & Ryan, Constance A. (1985). *Math problem solving: beginners through grade 3 .* Boston, MA:Allyn and Bacon.

242. Overholt, James L., Rincon, Jane B. & Ryan, Constance A. (1984). *Math problem solving for grades 4 through 8 .* Boston, MA:Allyn and Bacon.

243. Paris, S. G., Calfee, R. C., Filby, N., Hiebert, E.F., Pearson, P.D., Valencia, S.W. & Wolf, K. P. (1992). A framework for authentic assessment . *The Reading Teacher, 46(2),* 88-98.

244. Park, Hee-Seo (1980). *Sensitivity to social situation : the development of children's social inference and the relationships among social inference, social problem-solving, and social competence.*

245. Parrill-Burnstein, Melinda (1981). *Problem solving and learning disabilitie : an information processing approach.* New York: Grune and Stratton.

246. Peak, Lois (1996). *Pursuing excellence - A study of U.S. Eighth-Grade Mathematics and Science Teaching, Learning, Curriculum, and Achievement in International Context.* Washington.DC: U.S. Department of Education, National Center for Educational statistics. Government Printing Office.

247. Pearl, Judea (1984). *Heuristics: intelligent search strategies for computer problem solving.* Reading, MA: Addison-Wesley Pub. Co.

248. Pelton, Joseph (1992). *Future View - Communications Technology and Society in the 21st Century.* New York: Baylin Publishing.

249. Perkins, David N. (1986). *Reasoning as It Is and Could Be. An Empirical Perspective.* Presented at Annual Conference of the American Educational Research Association.

250. Perraton, Hilary (Ed.) (1993). *Distance Education for Teacher Training.* NYC, New York: Routledge.

251. Perry, Walter (1977). *The Open University.* San Francisco, CA: Jossey-Bass.

252. Piaget, Jean (1997). *The Development of Thought: Elaboration of Cognitive Structures.* New York: Viking.

253. Podany, Zita (1990). *Software for Middle School Physical Science - A Critical Review of Products.* Portland, Oregon: MicroSIFT, Northwest Regional Educational Laboratory.

254. Polya, George (1984). *Collected papers.* Cambridge, Mass: MIT Press.

255. Polya, George (1957). *How to solve it; a new aspect of mathematical method. (2d ed.)* Garden City, N.Y. :Doubleday and Co. Inc.

256. Polya, George (1981). *Mathematical discovery : on understanding, learning, and teaching problem solving.* New York: Wiley.

257. Polya, George (1977). *Mathematical methods in science.* Washington, DC: Mathematical Association of America.

258. Polya, George (1954). *Mathematics and plausible reasoning.* Princeton: Princeton Univ. Pres.

259. Popham, W. J. (1993). Circumventing the high costs of authentic assessment . *Phi Delta Kappan, 74(6),* 470-473.

260. President's Committee of Advisors on Science and Technology (1997). *Panel on Educational Technology.* Washington, DC: Report to the President on the use of Technology to Strengthen K-12 Education in the United States.

261. Prince, George M. (1970). *The practice of creativity, a manual for dynamic group problem solving.* New York: Harper and Row.

262. Quilici, Alexander E. (1985). *Human problem understanding and advice giving : a computer model.*

263. Raebeck, Barry. (1998). *Transforming Middle Schools : A Guide to Whole-School Change. (2nd ed.).* Lancaster, PA: Technomic Pub. Co.

264. Raizen, S., Baron, J., Champagme, A., Hoertel, E., Mullis, I. & Oakes, J. (1990). *Assessment in Science Education: The Middle Years.* Washington, DC: The National Center for Improving Science Education, The Network, Andover, MA and the Biological Sciences Curriculum Study.

265. Reeves, F .W., Fansler, T., Houle, C. O. (1938). *Adult Education.* New York: McGraw Hill.

266. Reif, Fredreck (1986). Scientific Approaches to Science Education. *Physics Today, November.*

267. Report of the Wingspread group on higher education (1993). *An American imperative: Higher expectations for higher education.* The Johnson Foundation: Racine,WI.

268. Resnick, Daniel P. & Resnick,Lauren B. (1985). Standards, Curriculum, and Performance: A Critical and Comparative Perspective. *Educational Researcher,14,* 5-20.

269. Resnick, Lauren B. (1984). Comprehending and Learning: Implications for a Cognitive Theory of Instruction. In Mandl, H. , Stern, N. L. , Trabasso, T. (Eds.) *Learning and Comprehension of Text.* Hillsdale, NJ: Lawrence Erlbaum Associates Publishers.

270. Resnick, Lauren B. (1997). *Discourse, Tools, and Reasoning: Essays on Situated Cognition.* New York: Springer-Verlag.

271. Resnick, Lauren B. (1987). *Education and Learning to Think.* Washington, DC:National Academy Press.

272. Reusser, K. (1993). Tutoring systems and pedagogical theory: Representational tools for understanding, planning, and reflection in problem solving. In Lajoie, S.P. & Derry, S. J. (Eds.), *Computers as cognitive tools (*pp. 143-177). Hillsdale, NJ: Lawrence Erlbaum Associates Publishers.

273. Richardson, Jacques (Ed.) (1984). *Models of reality : shaping thought and action .* Mt. Airy, MD: Lomond Books.

274. Rickards, Tudor (1974). *Problem-solving through creative analysis.* New York: Wiley.

275. Riel, Margaret Mary (1982). *Computer problem-solving strategies and social skills of language-impaired and normal children.*

276. Riquarts, Kurt (1987). *Science and Technology Education and the Quality of Life:* Papers Submitted to the Fourth International Symposium on World Trends in Science & Technology Education. Kiel, Germany: Institute of Science Education at Kiel University.

277. Robertshaw, Joseph E., Mecca, Stephen J. & Rerick, Mark N. (1978). *Problem solving, a systems approach.* New York: Petrocelli Books.

278. Rogers, Jennifer & Groombridge, Brian (1976). *Right to Learn: The Case for Adult Equality.* London, UK: Arrow Books.

279. Rogoff, Barbara & Wertsch, James (Eds.) (1984). *Children's Learning in the Zone of Proximal Development.* San Francisco, CA: Jossey-Bass.

280. Ross, Shelagn & Scanlon,Eileen (1995). *Open Science: Distance Learning and Open Learning of Scientific Subjects.* London, UK: Chapman.

281. Rubinstein, Moshe F. (1986). *Tools for thinkingand problem solving.* Redwood City, CA: Prentice-Hall.

282. Sanders, Donald & Sanders, Judith (1984). *Teaching Creativity Through Metaphor: An Integrated Brain Approach.* New York: Longman.

283. Saxena, Meenakshi (1983). *Children: voyage into problem space.* New Delhi: Abhinav Publications.

284. Scandura, Joseph M. & Durnin, John, et al. (1977). *Problem solving: a structural/process approach with instructional implications .* New York: Academic Press.

285. Schmidt, Willian, McKnight, Curtis & Raizen, Senta (1997). *A Splintered Vision - An Investigation of U. S. Science and Mathematics Education.* Boston, MA: Kluwer Academic Publishers.

286. Schoenfeld, Alan H. (1983). *Problem solving in the mathematics curriculum: a report, recommendations, and an annotated bibliography.* Washington, DC: Mathematical Association of America, Committee on the Teaching of Undergraduate Mathematics.

287. Searle, John (1984). *Minds, Brains, and Science.* Cambridge, MA: Harvard University Press.

288. Servan-Schreiber, Jean-Jaques (1981). *The World Challenge.* New York: Simon and Schuster.

289. Sherwood, John J. & Hoylman,Florence M. (1977). *Utilizing human resources: individual versus group approaches to problem-solving and decision-making* . West Lafayette, IN: Institute for Research in the Behavioral, Economic, and Management Sciences, Krannert Graduate School of Management: Purdue University.

290. Silver, Edward A. (Ed.) (1985). *Teaching and learning mathematical problem solving: multiple research perspectives* . Hillsdale, NJ: Lawrence Erlbaum Associates Publishers.

291. Simon, Herbert Alexander (1979). *Models of thought.* New Haven: Yale University Press.

292. Slavin, Robert (1990). *Cooperative Learning: Theory, Research, and Practice.* Englewood Cliffs, NJ: Prentice Hall.

293. Smith, M. C & Pourchot, T. (1998). *Adult Learning and Development: Perspectives from Educational Psychology.* Hillsdale, NJ: Lawrence Erlbaum Associates Publishers.

294. Smith, Peter & Kelly, Mavis (Eds) (1987). *Distance Education and the Mainstream: Convergence in Education.* NYC, New York: Croom Helm.

295. Smith, Reid G & Arbor, Ann (1981). *A framework for distributed problem solving.* Mich: UMI Research Press.

296. Snow, Richard E. (Ed.), Federico, Pat-Anthony & Montague, William E. (1980). *Cognitive process analyses of learning and problem solving.* Hillsdale, NJ: Lawrence Erlbaum Associates Publishers.

297. Spender, Dale (1996). *Creativity and the Computer Education Industry.* Canberra: International Federation for Information Processing.

298. Stafford, Kenneth R. (1966). *Problem solving as a function of language; final report.* Tempe,Ariz.: Dept. of Counseling and Educational Psychology. Arizona State University.

299. Stein, Morris Isaac (1974). *Stimulating creativity* . New York: Academic Press.

300. Steinberg, Esther (1991). *Computer-Assisted Instruction - A Synthesis of Theory, Practive, and Technology.* Hillsdale, NJ: Lawrence Erlbaum Associates Publishers.

301. Stephenson, N. (1995). *The Diamond Age.* N. Y.: Bantam Spectra.

302. Sternberg, Robert (1996). *Successful Intelligence.* New York: Simon and Schuster.

303. Steward, Elizabeth Petrick (1995). *Beginning Writers in the Zone of Proximal Development.* Hillsdale, NJ: Lawrence Erlbaum Associates Publishers.

304. Swahn, Anders Lennart (2000). *A new Possibility of Education for All: The Design of a Learning System for the 21st Century.* Private communication.

305. Tam, Maureen (2000). Contructivism, Instructional Design, and Technology: Implications for Transforming Distance Learning. *Educational Techology and Society, 3.*

306. The Children's Partnership (2000). Online Content for Low-Income and Undeserved Americans: The Digital Divide's New Frontier. The Children's Partnership, available at: http://www.childrenspartnership.org/pub/low_income/

307. Tondl, Ladislav (1973). *Scientific procedures, a contribution concerning the methodological problems of scientific concepts and scientific explanatio* , Translated from the Czech by Short, David. Boston: D. Reidel Pub. Co.

308. Totten, S., Sills-Briegel, T., Barta, K., Digby, A. & Nielsen, W. (1996). *Middle Level Education. An Annotated Bibliography.* Westport, Conn.: Greenwood Press.

309. Tuma, David & Reif,Frederick (1980). *Problem Solving and Education: Issues in Teaching and Research.* Hillsdale, NJ: Lawrence Erlbaum Associates Publishers.

310. Twigg, C. (1995). The value of independent study. *EDUCOM Review, July-August.*

311. Ulschak, Francis L., Nathanson, Leslie & Gillan, Peter G. (1981). *Small group problem solving: an aid to organizational effectiveness.* Reading, MA: Addison-Wesley Pub. Co.

312. UNESCO (1983). *Science and Technology Education and National Development.* Paris: UNESCO.

313. United States Department of Education (1997). *Attaining Excellence: A TIMSS Resource Kit.* Washington, DC: United States Department of Education, available at: http:\\www.webcommission.org

314. United States Department of Education (2000). *The Power of the Internet for Learning - Moving from promise to practice.* Washington, DC:Web-Based Education Commission, United States Department of Education, available at: http://nces.ed.gov/timss/timss95/rsrc_pub.asp

315. Valencia, S. W. (1991). You carn't have authentic assessment without authentic content. *The Reading Teacher, 44(8),* 590-591.

316. Van der Veer, Rene & Valsiner, Jaan (1994). *The Vygotsky Reader.* Oxford, UK: Blackwell.

317. Van der Veer, Rene & Valsiner,Jaan (1991). *Understanding Vygotsky: a Quest for Synthesis.* Oxford, UK: Blackwell.

318. VanGundy, Arthur B. (1984). *Managing group creativity: a modular approach to problem solving.* New York: American Management Associations.

319. VanGundy, Arthur B. (1988). *Techniques of structured problem solving (2nd ed.).* New York: Van Nostr& Reinhold Co.

320. Victor, Edward & Kellough, Richard (1997). *Science for Elementary and Middle School.* Englewood Cliffs, NJ: Prentice Hall.

321. Vygotsky, Lev (1978). *Mind in Society,* edited by Cole, M. ,John-Steiner, V. , Scribner, S. , Souberman, E. Cambridge, MA: Harvard University Press.

322. Vygotsky, Lev (1987). *The Collected Works of L. S. Vygotsky.* Plenum Press, available at: http://www.bestpraceduc.org/people/LevVygotsky.html

323. Vygotsky, Lev (1926). *Thought and Language.* Cambridge, Mass: MIT Press.

324. Wagner, Richard K. & Sternberg, Robert J. (1984). Alternate Conceptions of Intelligence and their Implications for Education. *Review of Educational Research, 54.*

325. Walley, Carl W, (Ed.) & Gerrick, W. Gregory (1999). *Affirming Middle Grades Education Boston.* Boston, MA: Allyn and Bacon.

326. Weaver, Gabriela (2000). An Examination of the National Educational Longitudinal Study (NELS:88). Database to Probe the Correlation between Computer Use in Scchools and Improvement in Test Scores. *Journal of Science Education and Technology, 9.*

327. Weinstock, Harold & Bork, Alfred (1986). *Designing Computer-Based Learning Material.* Proceedings of San Miniato,Italy, NATO ASI Workshop. Berlin, Germany: Springer-Verlag.

328. Wells, David (1988). *Hidden connections; double meanings.* Cambridge [England], New York: Cambridge University Press.

329. Whimbey, Arthur (1982). *Problem solving and comprehension.* Philadelpha, PA: Franklin Institute Press.

330. Whimbley, Arthur & Lochhead,Jack (1984). *Beyond Problem Solving and Comprehension - An Exploration of Quantitative Reasoning.* Philadelpha, PA: Franklin Institute Press.

331. Wickelgren, Wayne A. (1974). *How to solve problems; elements of a theory of problems and problem solving.* San Francisco:W. H. Freeman.

332. Wolff, Laurence (2000). Lifelong Learning for the Third Age. *TechKnowLogia, September/October.*

333. Zdenek, Marilee (1983). *The right-brain experience: an intimate program to free the powers of your imagination.* New York: McGraw-Hill.

334. Zelniker, Tamar & Jeffrey, Wendell E. (1977). *Reflective and impulsive children: strategies of information processing underlying differences in problem solving.* Chicago: University of Chicago Press.

335. Zinser. Jana (1992). *Project 2061: Education for a Changing Future.* Legislative Report, National Conference of State Legislatures. *17(17).*

INDEX